| | | |
|---|---|---|
| Vol. | 29. | **The Analytical Chemistry of Sulfur and Its Compounds** (*in three parts*). By J. H Karchmer |
| Vol. | 30. | **Ultramicro Elemental Analysis.** By Günther Tölg |
| Vol. | 31. | **Photometric Organic Analysis** (*in two parts*). By Eugene Sawicki |
| Vol. | 32. | **Determination of Organic Compounds: Methods and Procedures.** By Frederick T. Weiss |
| Vol. | 33. | **Masking and Demasking of Chemical Reactions.** By D. D. Perrin |
| Vol. | 34. | **Neutron Activation Analysis.** By D. De Soete, R. Gijbels, and J. Hoste |
| Vol. | 35. | **Laser Raman Spectroscopy.** By Marvin C. Tobin |
| Vol. | 36. | **Emission Spectrochemical Analysis.** By Morris Slavin |
| Vol. | 37. | **Analytical Chemistry of Phosphorus Compounds.** Edited by M. Halmann |
| Vol. | 38. | **Luminescence Spectrometry in Analytical Chemistry.** By J. D. Winefordner, S. G. Schulman and T. C. O'Haver |
| Vol. | 39. | **Activation Analysis with Neutron Generators.** By Sam S. Nargolwalla and Edwin P. Przybylowicz |
| Vol. | 40. | **Determination of Gaseous Elements in Metals.** Edited by Lynn L. Lewis, Laben M. Melnick, and Ben D. Holt |
| Vol. | 41. | **Analysis of Silicones.** Edited by A. Lee Smith |
| Vol. | 42. | **Foundations of Ultracentrifugal Analysis.** By H. Fujita |
| Vol. | 43. | **Chemical Infrared Fourier Transform Spectroscopy.** By Peter R. Griffiths |
| Vol. | 44. | **Microscale Manipulations in Chemistry.** By T. S. Ma and V. Horak |
| Vol. | 45. | **Thermometric Titrations.** By J. Barthel |
| Vol. | 46. | **Trace Analysis: Spectroscopic Methods for Elements.** Edited by J. D. Winefordner |
| Vol. | 47. | **Contamination Control in Trace Element Analysis.** By Morris Zief and James W. Mitchell |
| Vol. | 48. | **Analytical Applications of NMR.** By D. E. Leyden and R. H. Cox |
| Vol. | 49. | **Measurement of Dissolved Oxygen.** By Michael L. Hitchman |
| Vol. | 50. | **Analytical Laser Spectroscopy.** Edited by Nicolo Omenetto |
| Vol. | 51. | **Trace Element Analysis of Geological Materials.** By Roger D. Reeves and Robert R. Brooks |
| Vol. | 52. | **Chemical Analysis by Microwave Rotational Spectroscopy.** By Ravi Varma and Lawrence W. Hrubesh |
| Vol. | 53. | **Information Theory As Applied to Chemical Analysis.** By Karel Eckschlager and Vladimir Štěpánek |
| Vol. | 54. | **Applied Infrared Spectroscopy: Fundamentals, Techniques, and Analytical Problem-solving.** By A. Lee Smith |
| Vol. | 55. | **Archaeological Chemistry.** By Zvi Goffer |
| Vol. | 56. | **Immobilized Enzymes in Analytical and Clinical Chemistry.** By P. W. Carr and L. D. Bowers |
| Vol. | 57. | **Photoacoustics and Photoacoustic Spectroscopy.** By Allan Rosencwaig |
| Vol. | 58. | **Analysis of Pesticide Residues.** Edited by H. Anson Moye |
| Vol. | 59. | **Affinity Chromatography.** By William H. Scouten |
| Vol. | 60. | **Quality Control in Analytical Chemistry.** By G. Kateman and F. W. Pijpers |
| Vol. | 61. | **Direct Characterization of Fineparticles.** By Brian H. Kaye |
| Vol. | 62. | **Flow Injection Analysis.** By J. Ruzicka and E. H. Hansen |

(*continued on back*)

# Fluorometric Analysis in
Biomedical Chemistry

# CHEMICAL ANALYSIS

## A SERIES OF MONOGRAPHS ON ANALYTICAL CHEMISTRY AND ITS APPLICATIONS

*Editors*
**J. D. WINEFORDNER**
*Editor Emeritus*: **I. M. KOLTHOFF**

*Advisory Board*

Fred W. Billmeyer, Jr.    Victor G. Mossotti
Eli Grushka              A. Lee Smith
Barry L. Karger          Bernard Tremillon
Viliam Krivan            T. S. West

**VOLUME 109**

A WILEY-INTERSCIENCE PUBLICATION

**JOHN WILEY & SONS, INC.**

New York / Chichester / Brisbane / Toronto / Singapore

# Fluorometric Analysis in Biomedical Chemistry

## Trends and Techniques Including HPLC Applications

**NORIO ICHINOSE**
Department of Chemistry
Hamamatsu University
Hamamatsu, Japan

**GEORGE SCHWEDT**
Clausthal University
West Germany

**FRANK MICHAEL SCHNEPEL**
Department of Food Chemistry
Stuttgart University
Stuttgart, Germany

**KYOKO ADACHI**
Department of Chemistry
Hamamatsu University
Hamamatsu, Japan

A WILEY-INTERSCIENCE PUBLICATION

**JOHN WILEY & SONS, INC.**

New York / Chichester / Brisbane / Toronto / Singapore

This English edition has been translated from the
original Japanese publication Modern Fluorometric
Analysis by N. Ichinose, G. Schwedt, F. M. Schnepel, and K. Adachi.

Copyright © Baifukan Co., Ltd., 1987

In recognition of the importance of preserving what has been
written, it is a policy of John Wiley & Sons, Inc., to have books
of enduring value published in the United States printed on
acid-free paper, and we exert our best efforts to that end.

Copyright © 1991 by John Wiley & Sons, Inc.

All rights reserved. Published simultaneously in Canada.

Reproduction or translation of any part of this work
beyond that permitted by Section 107 or 108 of the
1976 United States Copyright Act without the permission
of the copyright owner is unlawful. Requests for
permission or further information should be addressed to
the Permissions Department, John Wiley & Sons, Inc.

*Library of Congress Cataloging in Publication Data:*

Fluorometric analysis in biomedical chemistry: trends and techniques
including HPLC applications/Norio Ichinose . . . [et al.].
   p. cm.—(Chemical analysis; v. 109)
Translation from the Japanese.
"A Wiley-Interscience publication."
Includes bibliographical references.
ISBN 0-471-52258-9
1. Fluorimetry. 2. Biomolecules Analysis. 3. Clinical
chemistry. I. Ichinose, Norio. II. Series.
   [DNLM: 1. Chemistry, Clinical. 2. Chromatography, High Pressure
Liquid—methods. 3. Fluorometry. QD 79.F4 F646]
QP519.9.F58F58  1991
616.07'56—dc20
DNLM/DLC
for Library of Congress                                  90-12271
                                                                                             CIP

Printed in the United States of America

10 9 8 7 6 5 4 3 2 1

# PREFACE

Fluorometric analysis is an ultramodern analytical technique, serving as an alternative to radioassays, which have the highest sensitivity among the conventional analyses of the day for the determination of trace amounts of elements. Specifically, the fluorometric method as a novel analytic procedure in life sciences such as biochemistry and biomedical and clinical chemistry is becoming one of the most important applications of photoluminescence, since it offers several advantages over radioassays, that is, good selectivity, wide linear range, and absence of harm, also it offers high sensitivity equal to that obtained with radioassays.

This book deals with the basic principles and the newest applications of fluorometric analysis, particularly in the various fields of the above-mentioned life sciences over the last 10 years. Moreover, in adding new information, ideas, and philosophies to the material previously published on fluorometric analysis, this book's contribution is unique in three regards:

1. In previously published texts, there are no descriptions of the applications to the life sciences of fluorometric analysis using high-performance liquid chromatography (HPLC). In this book, however, many cases are described from recent literature on fluorometric analysis coupled with HPLC for the separation and determination of organic components in biological samples.

2. The chemical structures of fluorophors in the literature on determining components are precisely illustrated by diagrams in order to give the reader a correct understanding of the analytical principle involved.

3. The names of the chemical structures in figures, tables, protocols, and the text are given using IUPAC Nomenclature as much as possible. In addition, all dimensions in the book are expressed in terms of SI (Système International) units.

This book represents an international scientific study, conducted across national boundaries, on fluorometric analysis and contains the latest information and ideas on the life sciences from the East and West as described by the four authors (two from Japan and two from Germany). It is highly useful both as a theoretical and a practical analytical textbook, and as a handbook on fluorometric analysis for undergraduate students, graduate students, professors, and research workers at universities, national institutes, hospitals, and related institutions and organizations.

I dedicate this book to the memory of the late Emeritus Professor Dr. Hidehiro Gotô of Tohoku University, and the late Emeritus Professor Dr. Michio Ôta of Yamanashi University. Professor Gotô guided my main progress in analytical chemistry and introduced me to Emeritus Professor Dr. Oskar Glemser of Göttingen University, the former president of the Chemical Society of Germany (F.R.G.), in 1979. While studying at Göttingen University under Professor Glemser, I became acquainted with two excellent analytical chemists, Professor Dr. Georg Schwedt and Dr. F.-M. Schnepel. This fortunate meeting led to a long and happy cooperation, which has resulted in the present study.

I am particularly grateful to Emeritus Professor Dr. Oskar Glemser of Göttingen University, Emeritus Professor Dr. Taichiro Fujinaga of Kyoto University, Emeritus Professor Dr. Shigero Ikeda of Osaka University, Emeritus Professor Dr. K.-G. Bergner of Stuttgart University, Dr. Yoshiro Kawashima, president, Dr. Martha Lee Alexander, and Mrs. Mina Toyama of Hamamatsu University School of Medicine, for their guidance and encouragement. Fluent in Japanese, Dr. Stephen G. Schulman of the University of Florida was particularly helpful in critically reading the completed translation to insure an accurate rendering of our science into English consistent with Wiley's *Chemical Analysis Series*. And I would like to express my special thanks to Mr. Tsuyoshi Nohara, Director, Editorial Department, and Mr. Kiichiro Honma of Baifukan Publishing Company in Japan; and to Mrs. Lauren C. Fransen, Director, International Rights, and Mr. James L. Smith, Senior Editor of John Wiley & Sons, for giving us the opportunity to publish this book.

It is a great pleasure for us to see this book become a reality. We hope it will contribute much to the progress of analytical chemistry and biomedical chemistry, as well as to the advancement of international scientific exchange and cooperation.

NORIO ICHINOSE

*On behalf of the authors, at Lake Hamana in Japan, Summer 1989.*

# CONTENTS

**CHAPTER 1  INTRODUCTION**     1
G. Schwedt


**CHAPTER 2  PHYSICAL PROPERTY OF FLUORESCENCE**     5
N. Ichinose

1. Photoluminescence     5
2. Photoabsorption and Emission     6
3. Inherent Characteristics of Fluorescence Emission     12
   3.1. Stokes' Shift     12
   3.2. Kasha's Rule     12
   3.3. Mirror-Image Rule     13
4. Fluorescence Quantum Yield     15
5. Influence of Various Factors on Fluorescence in Solutions     15
   5.1. Solvents     15
   5.2. Temperature     16
   5.3. pH     16
   5.4. Quenching     16
6. Fluorescence and Chemical Structure of Molecules     17
   6.1. Planar Structure     18
   6.2. Conjugated Double Bond and Resonance Structure     18
7. Fluorophors     21
   7.1. Intrinsic Fluorophors     21
     *7.1.1. Proteins*     22
     *7.1.2. Vitamin $B_2$*     22
     *7.1.3. NADH and $NAD^+$*     22

|  |  |  |
|---|---|---|
|  | 7.2. Extrinsic Fluorophors | 24 |
|  |    7.2.1. Fluorescein and Rhodamine Isocyanates and Isothiocyanates and Dansyl Chloride (DNS–Cl) | 24 |
|  |    7.2.2. Extrinsic Fluorophors for Carboxylic Acids | 24 |
|  | 8. Basis of Luminescence Analysis | 29 |
|  | 8.1. The Place of Luminescence Assays in Biochemistry | 29 |
|  | 8.2. Terminology and Definitions | 30 |
|  | 8.3. Historical Aspect | 30 |
|  | 8.4. Types of Luminescence | 31 |
|  |    8.4.1. Chemiluminescence | 31 |
|  |    8.4.2. Bioluminescence | 34 |
|  | References | 44 |
| **CHAPTER 3** | **PRINCIPLE OF FLUORESCENCE MEASUREMENTS** | **51** |
|  | *N. Ichinose and G. Schwedt* |  |
|  | 1. Apparatus and Arrangements | 51 |
|  | 1.1. Source of Light | 53 |
|  | 1.2. Monochromator and Filters | 56 |
|  | 1.3. Measuring Cells | 58 |
|  | 1.4. Photomultiplier | 60 |
|  | 2. Fluorescence Measurements | 62 |
|  | 2.1. Fluorescence Emission Spectrum | 62 |
|  | 2.2. Fluorescence Excitation Spectrum | 64 |
|  | 2.3. Calibration Curve of Fluorescence | 65 |
|  | References | 66 |
| **CHAPTER 4** | **BIOCHEMICAL AND BIOMEDICAL APPLICATIONS** | **69** |
|  | 1. Biochemistry | 69 |
|  | *F. M. Schnepel* |  |
|  | 1.1. Amines | 69 |
|  |    1.1.1. Native Fluorescence | 69 |
|  |    1.1.2. Derivatizations | 70 |
|  | 1.2. Amino Acids and Imino Acids | 83 |
|  |    1.2.1. Native Fluorescence | 83 |
|  |    1.2.2. Derivatizations | 85 |

| | | |
|---|---|---|
| 1.3. | Alkaloids | 89 |
| | *1.3.1. Native Fluorescence* | 90 |
| | *1.3.2. Derivatizations* | 94 |
| 1.4. | Vitamins and Related Compounds | 94 |
| | *1.4.1. Native Fluorescence* | 97 |
| | *1.4.2. Derivatizations* | 98 |
| 1.5. | Steroids | 101 |
| | *1.5.1. Estrogens* | 101 |
| | *1.5.2. Cholesterol* | 103 |
| | *1.5.3. Other Steroids* | 104 |

2. Biomedical and Clinical Chemistry — 106
N. Ichinose and K. Adachi

- 2.1. Biomedical Chemistry — 107
  - *2.1.1. Autoanalysis by Fluorometry of Enzyme Estimation of Free and Total Cholesterol* — 107
  - *2.1.2. Fluorometric Determination of Ammonia in Protein-Free Filtrates of Human Blood Plasma* — 107
  - *2.1.3. Fluorometric Determination of Tetracyclines in Biological Materials* — 108
  - *2.1.4. Erythrocyte Uroporphyrinogen I Synthase Activity in Diagnosis of Acute Intermittent Porphyria* — 109
  - *2.1.5. Quantitative Analysis of Bile Acids and Their Conjugates in Duodenal Aspirate by Fluorometry after Cellulose Acetate Electrophoresis* — 110
  - *2.1.6. Enzymatic Spectrofluorometric Determination of Uric Acid in Microsamples of Plasma by Using p-Hydroxyphenylacetic Acid as a Fluorophor* — 110
  - *2.1.7. Simplified Luciferase Assay of $NAD^+$ Applied to Microsamples from Liver, Kidney, and Pancreatic Islets* — 112
  - *2.1.8. Automated Fluorometric Analysis of Galactose in Blood* — 112
  - *2.1.9. Evaluation of Fluorometrically Estimated Serum Bile Acid in Liver Disease* — 113
  - *2.1.10. Fluorometric Determination of Plasma Unesterified Fatty Acid* — 114

2.1.11. New Fluorometric Method for Determination of Picomoles of Inorganic Phosphorus; Application to Renal Tubular Fluid     115
2.1.12. Assay for Nanogram Quantities of DNA in Cellular Homogenates     115
2.1.13. Intracellular pH Determination by Fluorescence Measurements     116
2.1.14. Detection of Diamagnetic Cation in Tissue Using the Fluorescent Probe Chlorotetracycline     117
2.1.15. Fluorometric Microassay of DNA Using a Modified Thiobarbituric Acid Assay     118
2.1.16. Fluorometric Oxidase Assays: Pitfalls Caused by the Action of Ultraviolet Light on Lipids     119
2.1.17. Fluorometric Analysis: A Study on Fluorescent Indicators for Measuring Near-Neutral ("Physiological") pH Values     120
2.1.18. Prolidase Deficiency: Characteristics of Human Skin Fibroblast Prolidase Using Colorimetric and Fluorometric Assays     120

2.2. Immunology     121

2.2.1. Types of Fluoroimmunoassay     122
2.2.2. Laser Fluorescence Immunoassay of Insulin     125
2.2.3. Luminescence Immunoassay of Human Serum Albumin with Hemin as Labeling Catalyst     126
2.2.4. Chemiluminescence-Labeled Antibodies and Their Applications in Immunoassays     127
2.2.5. Acridinium Esters as High-Specific-Activity Labels in Immunoassay     128
2.2.6. Two-Site Immunochemiluminometric Assay for Human $\alpha_1$-Fetoprotein     129
2.2.7. Homogeneous Immunoassay Based on Chemiluminescence Energy Transfer     129

2.2.8. Enhanced Luminescence Procedure for Sensitive Determination of Peroxidase-Labeled Conjugates in Immunoassay … 130
2.2.9. Chemiluminescent Tags in Immunoassays … 131
2.2.10. Direct Solid-Phase Fluoroenzymeimmunoassay of 5β-Pregnane-3α,20α-diol-3α-glucuronide in Urine … 133
2.2.11. On-Line Computer Analysis of Chemiluminescent Reactions, with Application to a Luminescent Immunoassay for Free Cortisol in Urine … 133
2.2.12. Solid-Phase Chemiluminescence Immunoassay for Progesterone in Unextracted Serum … 134

2.3. Medical Diagnoses … 135

2.3.1. Fluorometric Determination of δ-Aminolevulinate Dehydratase Activity in Human Erythrocytes as an Index of Lead Exposure … 135
2.3.2. Fluorescent Assay of Total Serum Cholesterol, with Use of Gas-Liquid Chromatography to Study Saponification Efficiency … 136
2.3.3. New Fluorometric Analysis for Mandelic and Phenylglyoxylic Acids in Urine as an Index of Styrene Exposure … 137
2.3.4. Fluorescence Methods in the Diagnosis and Management of Diseases of Tetrapyrrole Metabolism … 138
2.3.5. Sensitive Fluorometry of Heat-Stable Alkaline Phosphatase (Regan Enzyme) Activity in Serum from Smokers and Nonsmokers … 138

2.4. Obstetrics and Gynecology … 139

2.4.1. New Rapid Assay of Estrogens in Pregnancy Urine Using the Substrate Native Fluorescence … 139
2.4.2. A Semiautomated Method for the Determination of Estrogens in Early Morning Urine Specimens from Normal and Infertile Women … 140

2.4.3. Semiautomated Fluorometric Method for the Determination of Total Estrogens in Pregnancy Urine — 142
 2.4.4. Evaluation of an Aqueous Flourometric Continuous-Flow Method for Measurement of Total Urinary Estrogens — 142
2.5. Other Clinical Fields — 142
 2.5.1. Inherent Fingerprint Luminescence–Detection by Laser — 142
 2.5.2. New Dual-Staining Technique for Simultaneous-Flow Cytometric DNA Analysis of Living and Dead Cells — 143
 2.5.3. Continuous Measurement of Concentrations of Alcohol Using a Fluorescence-Photometric Enzymatic Method — 144
References — 146

## CHAPTER 5  BIOCHEMICAL AND BIOMEDICAL APPLICATIONS OF FLUOROMETRIC ANALYSIS USING HPLC — 159

N. Ichinose, F.-M. Schnepel, G. Schwedt, and K. Adachi

1. Biochemistry — 159
   1.1. Amines, Amino Acids, and Related Compounds — 159
     1.1.1. Amines (General) — 159
     1.1.2. N-Heterocyclic Compounds — 160
     1.1.3. Polyamines — 163
     1.1.4. Amino Sugars — 164
     1.1.5. Amino Acids and Imino Acids — 165
     1.1.6. Some Important Applications — 165
   1.2. Vitamins — 172
   1.3. Steroids — 175
     1.3.1. Bile Acids — 175
     1.3.2. Other Steroids — 177
   1.4. Organic Acids — 179
   1.5. Other Compounds — 181
     1.5.1. Alkaloids — 181
     1.5.2. Thiols — 182

2. Biomedical and Clinical Chemistry and
   Fisheries Biochemistry    183
   2.1. Cerebrospinal Fluid Polyamine Monitoring in Central Nervous System Leukemia    186
   2.2. Fluorescent Derivatives of Prostaglandins and Thromboxanes for Liquid Chromatography    186
   2.3. Use of Native Fluorescence Measurements and the Stopped-Flow Scanning Technique in the HPLC Analysis of Catecholamines and Related Compounds    188
   2.4. Determination of Plasma and Urinary Cortisol by HPLC Using Fluorescence Derivatization with Dansyl Hydrazine    189
   2.5. Determination of Urinary Placental Estriol by Reversed-Phase HPLC with Fluorescence Detection    189
   2.6. Simple and Rapid Determination of 5-Hydroxyindole-3-Acetic Acid in Urine by Direct Injection on a Liquid Chromatographic Column    189
   2.7. Simple Method for the Assay of Urinary Metanephrines Using HPLC with Fluorescence Detection    190
   2.8. Determination of the $B_2$ Vitamer Flavin-Adenine Dinucleotide in Whole Blood by HPLC with Fluorometric Detection    192
   2.9. Liquid-Chromatograph Profiling of Endogenous Fluorescent Substances in Sera and Urine of Uremic and Normal Subjects    192
   2.10. Liquid-Chromatographic Study of Fluorescent Materials in Uremic Fluids    193
   2.11. Determination of Serum Bile Acids by HPLC with Fluorescence Labeling    193
   2.12. Fluorescence HPLC of Eicosapentaenoic Acid in Serum and Whole Blood of Fish and in Body Fluid of Plankton after Labeling with 9-Anthryldiazomethane    196

2.13. Improved Assay of Unconjugated Estriol in Maternal Serum or Plasma by Adsorption and Liquid Chromatography with Fluorometric Detection ... 198
2.14. Improved Determination of Estriol-16α-Glucuronide in Pregnancy Urine by Direct Liquid Chromatography with Fluorescence Detection ... 199
2.15. Liquid-Chromatographic Study of Fluorescent Compounds in Hemodialysate Solutions ... 199
2.16. Determination of $B_2$ Vitamers in Serum of Fish Using HPLC with Fluorescence Detection ... 200
3. Food Chemistry ... 201
References ... 205

**INDEX** ... **211**

# CHAPTER 1

# INTRODUCTION

## G. SCHWEDT

Fluorometry is superior to photometry in the area of analytical spectroscopy because of its high sensitivity and selectivity. Many substances are, in principle, amenable to photometric analysis since they absorb light in the ultraviolet/visible region. In contrast, inherently fluorescent substances are few so that fluorometry based on native fluorescence is limited. The spectral properties of these substances are characterized by both excitation and emission spectra as well as by other fluorescent properties (e.g., fluorescence quantum yields), which are remarkably influenced by the environment (especially the solvent) and the structural parameters.

The apparatus for the analytical utilization of the fluorescent properties of chemical substances is no more expensive than that for spectrophotometry. Fluorometric measurements are carried out either in static solutions, as in photometry in a cuvette, or in combination with liquid chromatography (mainly high-performance liquid chromatography, HPLC) by using a flow cell. Fluorometry, like the alternative detection methods of HPLC, is no more subject to interferences by contamination than are the cuvette methods, so that the former is becoming more and more important. Moreover, it has the advantage that the simultaneous determination of fluorescent substances is possible. The principal factor controlling the fluorescence yield in HPLC is the mobile phase (the solvent of fluorescence measurement). On the other hand, HPLC can remove coexistent nonfluorescent substances that diminish the intensity of fluorescence. Therefore, in most cases, the accuracy of fluorometric analyses is improved by the use of HPLC. The exogenous interferences will be eliminated on the basis of the automatic processes of separation and continuous measurements. Thus, the improvement of reproducibility in the fluorometric measurements can be achieved.

In biochemical analysis, some aspects are particularly important for the fluorometric analysis or technique. For complex matrices such as urine, blood, the other body fluids, and various tissues, it is necessary to isolate a desired substance. In this case, care must be taken that contaminations do not result from the other fluorescent substances. On the other hand, the substances present that diminish fluorescence strength or the other fluorescent substances should be removed by the precleanup. The effort required for this process is slight because the selectivity in fluorometry is higher than that in spectrophotometry.

As for the trace analysis of organic substances of biochemical or medical interest, nonfluorescent substances can be transformed into fluorescent derivatives. Recently, new reagents have been developed, which are suitable both for group determinations and for analysis of an individual substance, for example, after a selective multistage reaction. For three reasons, these derivatization reactions play an especially important role in fluorometric analysis. First, the chemical derivatization can be carried out directly in a solution for measurement in a cuvette. Second, in group derivatizations particularly, mixed substances that can be separated by liquid chromatography and detected fluorometrically are obtained. Third, the substances that can be separated well by liquid chromatography are transformed into fluorescent derivatives after separation. Precolumn and postcolumn derivatization methods, as mentioned above, have been important in the development of fluorometric analysis in the last 10 years. Postcolumn derivatizations in so-called reaction detectors require only a little additional apparatus. Precolumn derivatization exhibits the further advantage that in many cases, better separation of the derivatives is possible than with the original substances.

For the effective application of these different fluorometric methods with or without HPLC, it is always necessary to consider carefully the specific peculiarities of fluorospectroscopy. Therefore the physical property of fluorescence is treated in detail in Chapter 2, where the main emphases are placed on understanding for the optimization of fluorometric measurement and the avoidance of interference. This is necessary because the fluorometric method is more susceptible to interference than the photometric method.

The essential equipment for fluorometric analysis is comparable to that for photometric analysis, and its basis is treated in Chapter 3. Biochemical and biomedical applications of fluorometry may be divided into two broad categories; in which HPLC is not employed (Chapter 4), and in which separation by HPLC is carried out prior to fluorometric detection (Chapter 5).

Important substances and their derivatizations for fluorometric analysis are discussed mainly in Chapter 4. Some significant points for the applications involve substances such as amines, amino acids, vitamins, steroids, organic acids, and alkaloids in the fields of biochemistry and biomedical and clinical chemistry. There, a delineation between "biochemistry" and "biomedical and clinical chemistry" is not always possible. In pharmaceutical, food, and agricultural chemical analyses, examples of application are clearly few. However, in these fields, further developments are expected, particularly in combination with HPLC.

Moreover, the advance of fluorometric analysis into the region of lower concentration may be achieved with the apparatus of laser fluorometry.

# CHAPTER 2

# PHYSICAL PROPERTY OF FLUORESCENCE

## N. ICHINOSE

## 1. PHOTOLUMINESCENCE

Luminescence is the emission of photons of frequency $v$, which have energy without heat, corresponding to $E_2 - E_1 = hv$. It arises from molecular relaxation from electronically excited states of higher energy $E_2$ to ground state $E_1$, which occurs after a substance absorbs light, heat, X-rays, chemical energy, and so on. Photoluminescence, in particular, is emitted by molecules in going from the excited state to the ground state after a substance absorbs the energy of light. At this time, the photoabsorption and emission do not permit an arbitrary energy state. Rather, transition takes place only between quantum mechanically permissible stationary states of the molecule. Photoluminescence is of two types, fluorescence and phosphorescence. Fluorescence is an emission that results from the return to the ground electronic state from the lowest energy excited singlet state. Such electronic transition is quantum mechanically "allowed," and the emission disappears as soon as irradiation of light from outside ceases. The emission rate results in a fluorescence lifetime near $10^{-8}$ s. On the other hand, phosphorescence is an emission that results from the transition from the triplet excited state to the lowest singlet ground state. Such a transition is not allowed, and its emission rate is slower than that of the former. Typical phosphorescence lifetimes range from $10^{-3}$ s to several seconds, as shown in Fig. 2.1. The intensity of

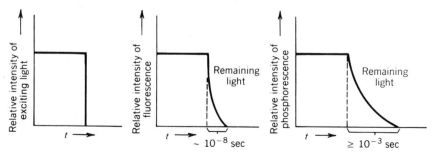

**Figure 2.1.** Time of remaining light in fluorescence and phosphorescence.

emission ($F$), after the irradiation by exciting light has ceased, decays according to Eq. (2.1):

$$F = F_0 e^{-t/\tau} \qquad (2.1)$$

where $F_0$ is the intensity of emission when a substance has just been excited, $t$ is the time after the irradiation by exciting light has ceased, and $\tau$ is the average lifetime of the excited state and is given by Eq. (2.2), where $\rho(t)$ represents the decay function:

$$\tau = \frac{\int_0^\infty t\rho(t)\,dt}{\int_0^\infty \rho(t)\,dt} \qquad (2.2)$$

Recently, fluorometric analysis has been widely applied to identification and determination in the various fields of biochemistry, especially biomedical and clinical chemistry, as a highly sensitive and selective technique. Phosphorescence is observed mainly in the solid state at extremely low temperature. Its sensitivity is inferior to that of fluorescence, but identification of the desired component is possible by measurements of its spectrum, intensity, and lifetime ($\tau$).

## 2. PHOTOABSORPTION AND EMISSION

Light has properties of both a wave and a particle (light quantum). According to Einstein (1), the energy of a light quantum $\varepsilon$ is given by Eq. (2.3):

$$\varepsilon = h\nu = \frac{hC}{\lambda} \qquad (2.3)$$

where $h$ is Planck's constant ($6.626176 \times 10^{-34}$ J.s.), $\nu$ is frequency ($s^{-1}$), $c$ is the velocity of light ($2.997925 \times 10^8$ m s$^{-1}$), and $\lambda$ is the wavelength of light (m). Then Eq. (2.3) can be rewritten as Eq. (2.4):

$$\varepsilon = \frac{6.626176 \times 10^{-34} \times 2.997925 \times 10^8}{\lambda} \text{ (J)} \qquad (2.4)$$

It seems reasonable to consider that the absorption and emission of light energy by a molecule, accompanying light excitation, result from gain and loss of the inner energies in the molecule, which consist of electronic excitation, vibration, and rotation energies. From the theory of Einstein in which "an atom or a molecule can absorb only a light quantum at a time," the light absorption energy $E$ per 1 mole of atom or molecule, that is, the photochemical equivalence, can be expressed as Eq. (2.5):

$$E = N_0 h\nu \qquad (2.5)$$

where $N_0$ is Avogadro's number ($6.022045 \times 10^{23}$ mol$^{-1}$).

It can be seen from the above-mentioned facts that, when a molecule causes photoexcitation from the ground state to higher energy states, the irradiation by ultraviolet rays of shorter wavelength is far more effective than that by infrared rays and visible rays. When a substance absorbs ultraviolet rays, an atom or a molecule of the substance changes to higher electron energy states from the ground state. In other words, the excited atom or molecule that absorbs the light having frequency $\nu$, corresponding to a difference in potential between the above two states, is able to form two electronic states, that is, singlet excited state $S$ and triplet excited state $T$, which bear the distinct spin configurations shown in Fig. 2.2.

The probability of electron transitions occurring between the ground and excited states on the occasion of light irradiation is proportional to the square of $\int \phi_i M \phi_j d\tau$, where $\phi_i$ and $\phi_j$ are the wave function of the ground and excited states, respectively, $M$ is the vector of electronic dipole moment, and $\tau$ is the fluorescence lifetime. If this probability has a finite value, the electronic transition is quantum mechanically "allowed"; otherwise, it is a "forbidden transition" that theoretically ought not to occur. Nevertheless, such a spin forbidden transition caused by asymmetric interatomic vibrations and interaction between spin and orbital angular momenta, as in excited state $T$ in Fig. 2.2, seldom actually arises. An electron can have a spin quantum number, $S$, of $\pm \frac{1}{2}$, and a normal polyatomic molecule in the ground state usually has an even number of electrons with paired

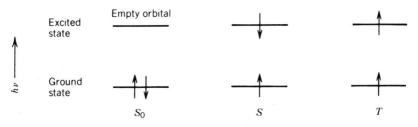

**Figure 2.2.** Excited state and spin.

spins; that is, there are as many electrons having $S = +\frac{1}{2}$ as those having $S = -\frac{1}{2}$. Therefore, the total spin of the molecule is zero.

Multiplicity ($m$), is a term used to express the spin angular momentum of a given electronic state in an atom or molecule; it is related to the electron spin through the expression $m = 2S + 1$. In a molecule having all electron spins paired, $m = 1$ because $S = \frac{1}{2} - \frac{1}{2} = 0$. Such a molecule is said to have a "singlet" electronic state. On the other hand, when the spin of a single electron in a polyatomic molecule is reversed through some internal energy transition, the molecule finds itself with two unpaired electrons, that is, $S = \frac{1}{2} + \frac{1}{2} = 1$, each electron occupying a different orbital. Consequently, $m = 3$; in this case the molecule is said to have a "triplet" electronic state. In Fig. 2.2, therefore, $S_0$ and $S$ represent singlet electronic states and $T$ is the triplet electronic state.

The electronic energy transitions occurring between photoabsorption and emission in a diatomic molecule are shown in Fig. 2.3 as plots of potential energy versus interatomic distance. The abscissa and the ordinate represent, respectively, the internuclear distance between atoms of $A$ and $B$ and the potential energy. The lowest potential energy level in the singlet ground state is represented as $E_{S_0}$; the higher potential energy levels in the singlet and triplet excited states are represented as $E_{S_1}$ and $E_{T_1}$, respectively; and the vibrational energy levels which have variously discontinuous energies in each state of $E_{S_0}$, $E_{S_1}$, and $E_{T_1}$ are shown as 0, 1, 2, 3, .... At ordinary temperatures it is assumed that before absorbing the light the molecule is in the lowest vibrational energy level of 0, that is, $A$–$B$, at the $E_{S_0}$ curve. When a quantum of light impinges on such a molecule and is absorbed by the molecule, electronic transition, due to the light absorption occurs almost instantaneously, in about $10^{-15}$ s, whereas the period of nuclear vibration at this time is about $10^{-12}$ s. In other words, this electronic transition is so rapid that the internuclear distance ($r$) can be considered as a remaining constant during the transition process [the Frank–Condon principle (2)]. It is possible to reason roughly, as follows, regarding the irradiation processes and return to the ground state:

1. In the case of a gaseous monoatom, the excited atom returns to the ground state, $E_{S_0}$, from the excited state $E_{S_1}$ and radiates light which has the same wavelength as that of the absorbed light because there are no vibrational energy levels between the excited and ground states. This radiation is termed *resonance radiation* or *resonance fluorescence*.

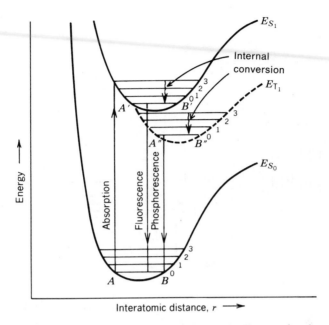

**Figure 2.3.** Plots of potential energy versus interatomic distance for the process of electronic energy transitions.

2. In the case of a diatomic molecule, after the molecule in a higher excited state relaxes radiationlessly, due to collision with other molecules of the solvent, to the lowest vibrational energy level of $A'$–$B'$ at the $E_{S_1}$ curve in Fig. 2.3, it persists there for a definite time (about $10^{-8}$ s), and then returns to the ground state with emission of light, that is, "fluorescence."

3. When the potential energy curves of $E_{S_1}$ and $E_{T_1}$ cross as shown in Fig. 2.3, the excited molecule is able to convert radiationlessly, in spite of the forbiddenness of the transition, to the excited triplet state from the excited singlet state, and transit further to the lowest vibration energy level of $A''$–$B''$ at the $E_{T_1}$ curve by relaxation. After remaining there for a relatively long time, (about $10^{-3}$ s), it returns to the ground state with emission of light, that is "phosphorescence."

Jablonski (3), Becker (4), Mataga and Kubota (5), Berlman (6), Zander (7), and Schwedt (8) have described in more detail the influences of the electronic and vibrational energy levels of a molecule on the processes of electronic transition as shown in Fig. 2.4. The ground, first, second, and third electronic energy levels in the singlet manifold

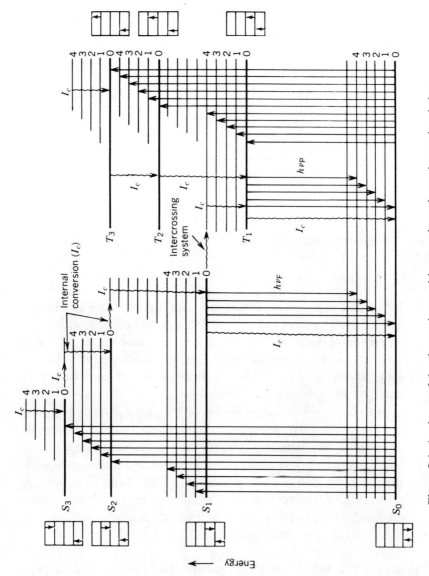

**Figure 2.4.** A scheme of the electronic transitions at photoabsorption and emissions.

and those except the ground state (there is none) in the triplet manifold are indicated by $S_0, S_1, S_2, S_3$, and $T_1, T_2, T_3$, respectively. In each of these potential energy levels, a molecule can exist in different vibrational energy levels numbered in order of increasing energy as 0, 1, 2, 3, and so on.

After photoabsorption, the excited molecule returns readily within $10^{-12}$ s, to the lowest vibrational energy level, 0, of state $S_1$ from the higher vibrational energy levels of state $S_1$ or above, by loss of the potential energy, because of the overlapping of vibrational energies of the highest vibrational energy levels of state $S_1$ or $S_2$ and the lowest vibrational energy level of state $S_2$ or $S_3$. Such a radiationless relaxation process is termed *internal conversion* (9, 10). After persisting in this lowest vibrational energy level of state $S_1$ for about $10^{-8}$ s, the molecule returns from there to the ground state resulting in fluorescence. Since the difference of potential energy between $S_0$ and $S_1$ is larger than that between $S_1$ and $S_2$ or $S_2$ and $S_3$, the above-mentioned internal conversion does not occur at the gap between $S_1$ and $S$. Generally internal conversion develops completely before fluorescence by reason that of the fact fluorescence lifetimes are typically near $10^{-8}$ s.

The excited molecule in the singlet state $S_1$ can also undergo conversion to the first triplet state $T_1$. Such conversion from $S_1$ to $T_1$ is called *intercrossing system* (11). The electron, after transiting further to the lowest vibration energy level, 0, of state $T_1$, by internal

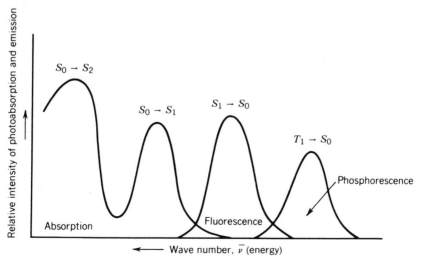

**Figure 2.5.** A scheme of photoabsorption and emission.

conversion and remaining there for a longer time (about $10^{-3}$ s or more), returns to the ground state with the emission of phosphorescence. The spectrum of phosphorescence generally occurs at a longer wavelength than that of fluorescence, as can be seen in Fig. 2.5.

## 3. INHERENT CHARACTERISTICS OF FLUORESCENCE EMISSION

### 3.1. Stokes' Shift

Except for atoms in the gas phase, the wavelength of the fluorescence emission spectrum invariably shifts to a lower energy range than that of the absorption spectrum by losing potential energy due to internal conversion and vibrational relaxation in state $S_1$ or above. Stokes (12) first observed this phenomenon in Cambridge in 1852 using a simple apparatus, as shown in Fig. 2.6, in which a source of UV rays was provided by sunlight and a blue glass filter, and the exciting sunlight was previously prevented from reaching the "eye fluorescence detector of Stokes" by a glass filled with yellow wine. Since quinine fluorescence occurs near 450 nm, visible color can be easily observed with the eye.

### 3.2. Kasha's Rule

Kasha (13) found that the emission of fluorescence or phosphorescence in organic compounds is usually independent of the excitation wavelength, that is, the excitation energy. In other words, in Fig. 2.4, even if a molecule is excited from state $S_0$ to a state higher than $S_1$, fluorescence or phosphorescence is emitted only from the lowest

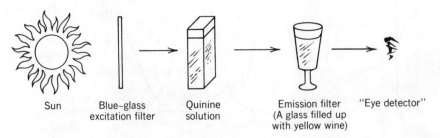

**Figure 2.6.** First observation of the Stokes' shift phenomenon.

vibrational energy level in state $S_1$ or $T_1$. This rule encompasses the fluorophors of organic compounds in the liquid and solid phases.

### 3.3. Mirror-Image Rule

Fluorescence spectra are classified into two categories: the fluorescence excitation spectrum, $E_x$, which indicates the relationship between wavelength of exciting light and fluorescence intensity at a constant wavelength of fluorescence, and the fluorescence emission spectrum, $E_m$, which indicates the relationship between fluorescence wavelength and its intensity at a constant wavelength of the exciting light.

Figure 2.7 is an example of absorption and fluorescence emission spectra taken from the plots of potential energy versus interatomic distance, in Fig. 2.3. When these spectra are plotted with the extinction coefficient as ordinate and the wave number ($\bar{v}$) as abscissa, the fluorescence emission spectrum appears to be a mirror image of the absorption spectrum, as shown in Fig. 2.8. The results in Figs. 2.7 and 2.8 do not contradict the previously described Stokes's and Kasha's rules.

Obviously, the three peaks of the absorption spectrum in Fig. 2.8 correspond to the electronic transitions from the lowest vibration level

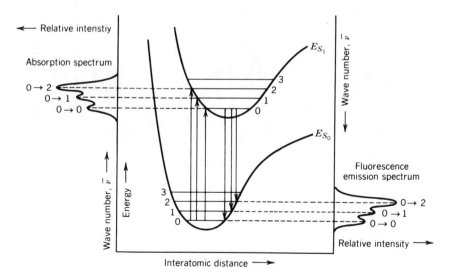

**Figure 2.7.** Analysis of absorption and fluorescence emission spectra from the plots of potential energy versus interatomic distance for the process of electronic energy transitions.

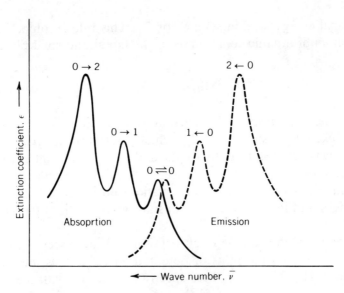

**Figure 2.8.** Relationship between the spectra of absorption and of fluorescence emission.

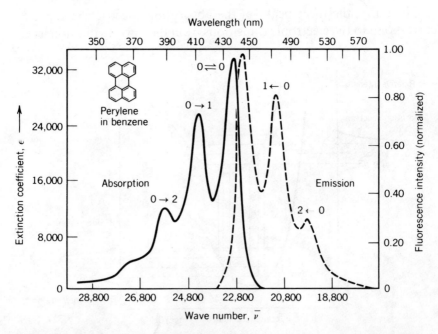

**Figure 2.9.** Absorption and fluorescence emission spectra of perylene. Reprinted from J. R. Lakowicz, *Principles of Fluorescence Spectroscopy*, Plenum Press, New York (1983).

of state $S_0$ to vibrational levels 0, 1, and 2 of state $S_1$ in Fig. 2.7, and the three peaks of the fluorescence emission spectrum also correspond to the transitions from the lowest vibration level of state $S_1$ to the vibration quantum numbers 0, 1, and 2 of state $S_0$. In other words, in Fig. 2.8, the three maxima of $(0 \rightarrow 0)$, $(0 \rightarrow 1)$, and $(0 \rightarrow 2)$ of the absorption spectrum correspond, respectively, to those of $(0 \leftarrow 0)$, $(1 \leftarrow 0)$, and $(2 \leftarrow 0)$ of the fluorescence emission spectrum.

In the fluorophors perylene, anthracene, fluorescein, rhodamine B, mirror images are also observed in both their spectra, as in the example of perylene in Fig. 2.9.

## 4. FLUORESCENCE QUANTUM YIELD

Fluorescence quantum yield, $Q_f$, is the ratio of the number of photons emitted to the number absorbed, and the ratio of fluorescence energy emitted to the energy absorbed is termed fluorescence energy yield, $\phi_f$. The fluorescence quantum yield can be written as follows.

$$Q_f = \frac{K_f}{K_f + K} \qquad (2.6)$$

where $K_f$ is the rate constant of fluorescence from a fluorophor, and $K$ is that of all radiationless transitions from $S_1$ to $S_0$ and from $S_1$ to $T_1$ in it. If $K$ is much smaller than $K_f$ (i.e., $K_f \gg K$), the fluorescence quantum yield can be close to unity, that is, $Q_f = 1$. However, the fluorescence energy yield is always less than unity.

## 5. INFLUENCE OF VARIOUS FACTORS ON FLUORESCENCE IN SOLUTIONS

### 5.1. Solvents

Generally solvents affect the fluorescence characteristics of substances greatly. As aspects of interaction between molecules of solute and solvent, there are the intermolecular transfer of an electron in a fluorophor due to polarity of the solvent, electrostatic action between dipole moments, dispersions, hydrogen bonds and so on. When the rates of internal conversion and intersystem crossing are affected by these factors, the fluorescence emission spectrum and its quantum yield are variable (4, 14).

## 5.2. Temperature

The higher the temperature in the solution rises, the lower fluorescence intensity becomes by reason (15, 16) of the increasing probability of intermolecular collision and the loss of potential energy due to the radiationless deactivation. As a practical problem accompanying a rise of temperature in the solution, the wavelength of the fluorescence emission spectrum shifts slightly, so it is preferable to measure the fluorescence intensity at lower temperature.

## 5.3. pH

Van Duuren (14) and Schulman (17) reported in detail about the dependence of fluorescence on pH. Generally, the variation of pH in solution has an influence on the fluorescence excitation and emission spectra at the time of transition of the proton.

## 5.4. Quenching

When a fluorophor coexists with some other substances, the fluorescence intensity falls remarkably because the fluorescence quantum yield decreases on account of radiationless deactivation by their mutual collision and other factors. Such phenomena are called *quenching* (18). When fluorophor and quencher are expressed as $F$ and $Q$, respectively, the quenching processes can be written as follows:

$F + h\nu_a \rightarrow F^*$ (Photoabsorption) ⎫ Photoexcitation
$F^* \rightarrow F + h\nu_f$ (Emission of fluorescence) ⎭ and emission
$F^* + Q \rightarrow F + Q$ (Quenching by foreign molecule) ⎫
$F^* + F \rightarrow F + F$ (Quenching by the same molecule) ⎭ Quenching processes

Principal examples of the quenching effect for fluorophores are described below.

1. Transition elements, particularly the colored metal ions such as $Cr^{3+}$, $Fe^{3+}$, $Ni^{2+}$, and $Cu^{2+}$ and anions like $Br^-$, $I^-$ and $NO_3^-$, are all quenchers of fluorescence.
2. Generally, the fluorescence intensity of a fluorophor in the gas or liquid state decreases with an increase in its concentration after

the fluorophor reaches above a certain level. This phenomenon is termed *concentration quenching*.
3. As described in Section 5.1, fluorescence intensity is variable according to the solvent's properties. Moreover, in solvents with a group as such C=C or C=O, the quenching effect is smaller than in solvents with C–N, C–Cl, or C–Br groups.

## 6. FLUORESCENCE AND CHEMICAL STRUCTURE OF MOLECULES

For a long time there has been research to find experimental rules for the relationship between the chemical structures of organic compounds and fluorescence (4, 19–23). For example, although neither phenolphthalein or pyridine has fluorescence, fluorescein, in which a bridge of –O– is added to the former, and quinoline, in which a benzene ring is added to the latter, have their respective fluorescences, as shown in Fig. 2.10. As a result of much research, it has been gradually discovered that as requisites for fluorescence in organic compounds, it is necessary to have high molar extinction coefficients, facile intramolecular electronic transfer at the time of photoabsorption and inefficient radiationless transitions in the process of return from the excited singlet state $S_1$ to the ground state $S_0$. The most important experimental results obtained to date are described below.

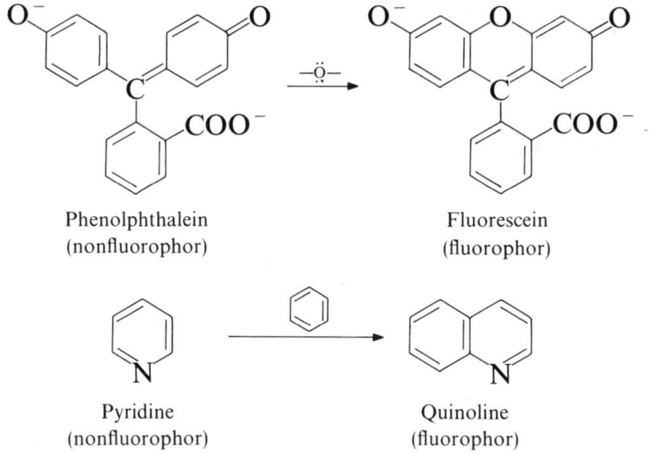

**Figure 2.10.** Relationship of fluorescence and chemical structure.

### 6.1. Planar Structure

As stated above, fluorescein, in which an oxygen bridge is added to phenolphthalein, has fluorescence. Eosin is also a natural fluorophor which has green fluorescence in alkaline solution and has a planar structure as well as fluorescein, as shown in Fig. 2.11.

Gotô (24) who perceived that fluorescein and eosin are intrinsic fluorophors, first applied their fluorescence characteristics to fluorometric analysis. Later, Förster (19) indicated that the planar structures of organic compounds have something close to do with their fluorescence phenomena, and Kasha (25) reported on the decrease of fluorescence intensity when a substituent which interrupts the planar structure is added to a cyanine dye. Further, Weller (26) and Hercules (27) explained, respectively, that a stilbene derivative of *trans* form has fluorescence, whereas the derivative of *cis* form has no fluorescence, and the reason is the difficulty of forming the plane structure in the case of the latter (Fig. 2.12).

### 6.2. Conjugated Double Bond and Resonance Structure

Yoshida and Oda (28) interpreted the relationship of the fluorescence phenomena of organic compounds to their chemical structures on the

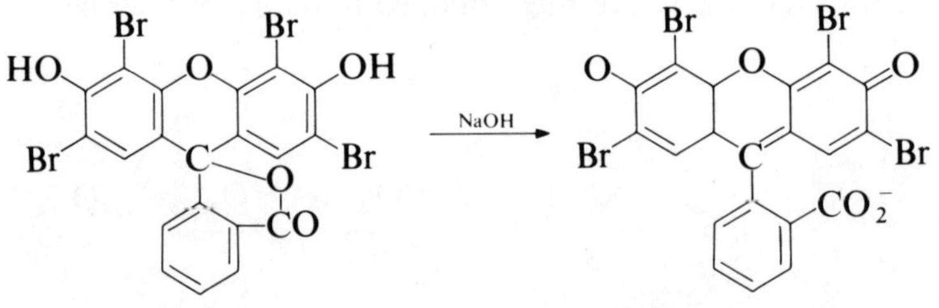


Eosin (nonfluorophor)

Eosin in alkai solution (fluorophor)

**Figure 2.11.** Plane structures of eosin.

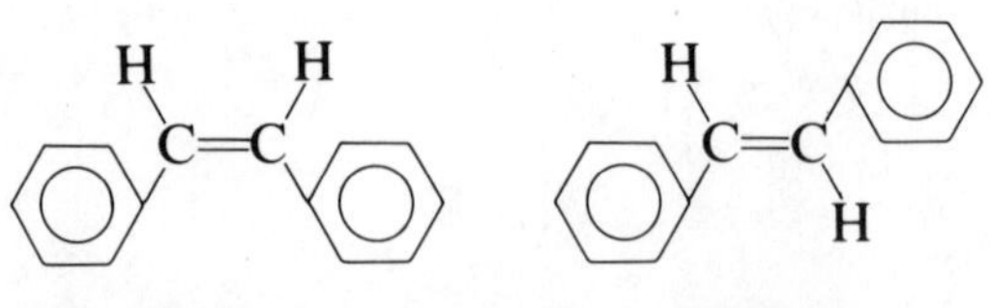

*cis*-form (nonfluorophor)

*trans*-form (fluorophor)

**Figure 2.12.** Isomers of stilbene.

basis of their experimental data as follows:

1. A molecule having resonance structures with conjugated double bonds, may demonstrate fluorescence. In fact, some fluorophors of condensed aromatic hydrocarbons have higher fluorescence intensities than those of chain compounds. Briefly, the larger the number of conjugated double bonds in the molecule, the higher the fluorescence intensity, as can be seen from an example in Table 2.1.
2. A molecule may fluoresce if it has a substituent that easily varies the electron density in a resonance structure with conjugated double bonds. In practice, when a substituent with a mesomeric effect (i.e., M effect) is introduced into the resonance structure in a molecule, its fluorescence intensity increases greatly. A group that causes a special increase in the photoabsorption is termed an auxochrome. Such an auxochrome can belong to one of two groups, $+ M_f$ and $- M_f$ and the electron transfer of the former occurs in the direction from substituent to benzene ring, while that of the latter take place in the opposite direction.

There are $-OR$, $-NH_2$, and so on, as $+ M_f$, and $-CN$, $CO$, and so on, as $- M_f$, where R is an H or alkyl group. The auxochromes $-OH$, $-OR$, or $-NH_2$ as $+ M_f$ have at least one lone electron pair in the atom adjacent, that is, "key atom" (11), to a benzene nucleus or other aromatic hydrocarbon. Figure 2.13 shows the resonance structures of phenol and aniline and the attached auxochrome of $-OH$ or $-NH_2$ as $+ M_f$. On account of the increasing electron density at the ortho and para positions in their compounds, both positions become targets for the attack of electrophilic reagents. On the other hand, benzonitrile in which a $-CN$ group is attached to a benzene nucleus, also has a resonance structure as shown in Fig. 2.14a; however, the

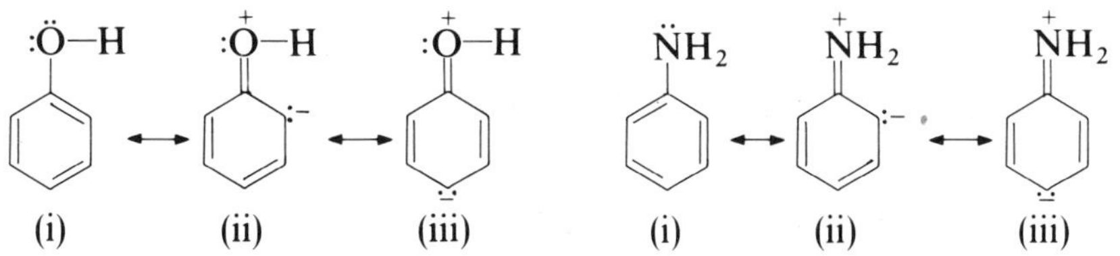

**Figure 2.13.** Resonance structures of phenol and aniline, attached auxochrome as $+ M_f$.

**Table 2.1. Fluorescence Characteristic of Some Condensed Aromatic Hydrocarbons**

| Substance | | Maximum Spectra [nm] | | Fluorescence Quantum Yield |
|---|---|---|---|---|
| | | $\lambda_{ex}$ | $\lambda_{em}$ | |
| Benzene | ⬡ | 205 | 278 | 0.11 |
| Naphthalene | ⬡⬡ | 286 | 321 | 0.29 |
| Anthracene | ⬡⬡⬡ | 365 | 400 | 0.46 |
| Tetracene | ⬡⬡⬡⬡ | 390 | 480 | 0.60 |

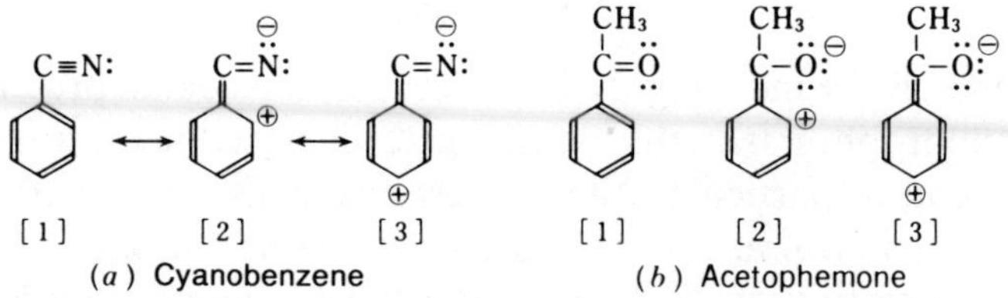

**Figure 2.14.** Resonance structures of cyanobenzene and acetophenone attached auxochrome as $-M_f$.

**Table 2.2. Effect of Different Substituents as Auxochromes on Fluorescence Intensity**

| Substituent | Effect on Fluorescence Intensity |
|---|---|
| $-OH$, $-OCH_3$, $-OC_2H_5$ | Increase |
| $-COOH$, $-CH_2-COOH$ | Increase |
| $-N(CH_3)_3^+$, $-NHCOCH$ | Remarkable reduce |
| $-SH$, $-F$, $-Cl$, $-Br$, $-I$ | Decrease |
| $-NH_2$, $-NHR$, $-NR_2$ | Increase |
| $-NO_2$, $-NO$ | Complete extinguishing |
| $-CN$ | Increase |
| $-SO_3H$ | No change |
| $=CO$, $-CHO$ | Decrease |
| Alkyl | Little increase or decrease |

electron density at the ortho and para positions accompanying the electron transfer from the benzene ring to the substituent $-CN$ decreases as compared with the case of phenol and aniline; thereby, both positions are susceptible to nucleophilic attack. Acetophenone, with the substituent C=O in Fig. 2.14b, is subject to a nucleophilic reaction similar to that of benzonitrile.

Table 2.2 shows the general effects of different substituents having $+M_f$ and $-M_f$ on fluorescence intensity.

## 7. FLUOROPHORS

### 7.1. Intrinsic Fluorophors

A number of biological molecules contain naturally occurring or intrinsic fluorophores. Some important ones are described below.

### 7.1.1. Proteins

Tryptophan is the most highly fluorescent amino acid in proteins. This natural fluorophor is highly sensitive to changes in its surrounding environment; in particular, the fluorescence emission spectrum shifts because of environmental changes such as the binding of ligands, and the association of indogenous and exogenous substances, and its fluorescence intensity generally is weaker in proteins.

### 7.1.2. Vitamin $B_2$

Vitamin $B_2$ is a generic term for the three naturally occurring flavin derivatives: riboflavin (RF), flavin mononucleotide (FMN), and flavin adenine dinucleotide (FAD). The latter two are phosphorylated, as shown in Fig. 2.15.

Vitamin $B_2$ occurs in various foods and is widely distributed, mainly as FAD, in leafy vegetables, meat, fish, eggs, milk, and milk products (29). In the intestine RF taken up with food is phosphorylated to FMN by the intestinal mucosa during absorption, and in tissue cells FMN can be further converted to FAD. FMN and FAD, which are called "yellow enzymes," combine with a certain protein and serve as coenzymes in most of the oxidation and reduction reactions catalyzed by the flavin enzyme. When these yellow enzymes (i.e., flavoproteins) are reduced by taking part in the oxidation–reduction reactions, they become colorless because two hydrogen atoms attach to the isoalloxazine nucleus of the flavin's coloring matter, as shown in Fig. 2.16. The flavin derivatives RF, FMN, and FAD all have yellow-green fluorescence, and their fluorescence emission spectra (30) are

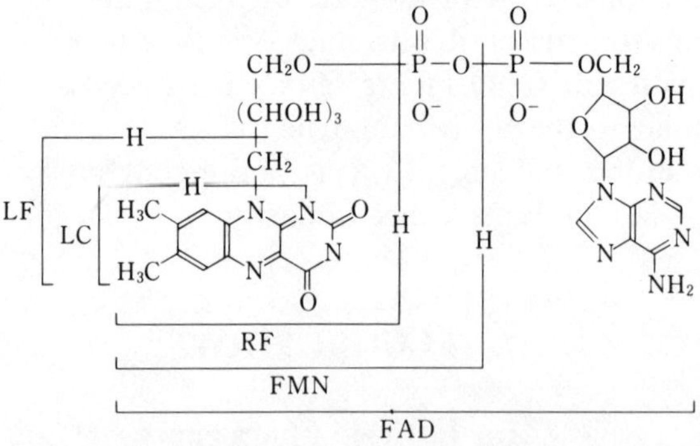

LF, lumiflavin; LC, lumichrome; RF, riboflavin;
FMN, flavinmononucleotide; FAD, flavinadeninedinucleotide.

**Figure 2.15.** Structures of $B_2$ vitamer flavins.

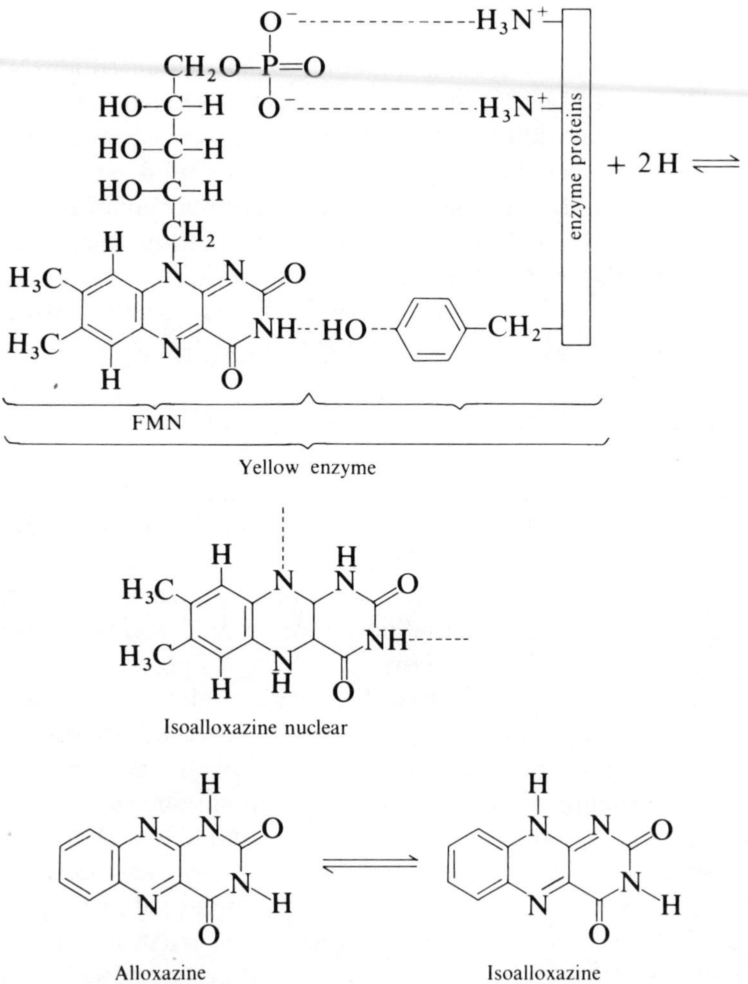

**Figure 2.16.** Structures of yellow enzymes and isoalloxazine nuclear of flavin's coloring matter.

similar to each other. However, the fluorescence quantum yields of RF and FMN are about 0.24, while the yield of FAD is considerable lower, 0.02, because of intramolecular quenching by adenine of the fluorophor of isoalloxazine. RF is unstable in aqueous solution against light and is decomposed to lumichrome (LC) in acid or neutral solution, and to lumiflavin (LF) in alkaline solution. The LC and LF photodecomposition products have fluorescence; in particular, LF is applied to the fluorometric analysis (31) of vitamin $B_2$ because of its high fluorescence intensity.

### 7.1.3. NADH and $NAD^+$

NADH is a reduced form of nicotinamide adenine dinucleotide (NAD) liberated as a coenzyme in biocells, and $NAD^+$ is the oxidizing form of it. NADH is a fluorophor of high fluorescence intensity, with absorption and fluorescence emission maxima at 340 and 450 nm, respectively, whereas $NAD^+$ has no fluorescence. The fluorophor of NADH is somewhat quenched by collision with the adenine nucleus, but when it is linked to a protein, the fluorescence intensity increases about fourfold because of the transformation in its chemical structure accompanying combination with protein.

## 7.2. Extrinsic Fluorophors

Many organic compounds of biochemical interest have no fluorescence; moreover, the fluorescence intensities of intrinsic fluorophores in organic macromolecules such as proteins are generally lower than those of extrinsic fluorophors. Recently, advanced techniques (32, 33) in which intrinsic nonfluorophors are covalently combined with certain extrinsic fluorophors to yield new fluorescent products, that is, labeling fluorophors, have been widely adopted to bioanalytical chemistry because the properties of extrinsic fluorophors can be suitably chosen for the desired experimentation; also, the sensitivity of labeling fluorophores for fluorometric determination is usually higher than that of many nominally intrinsic fluorophors. For example, such fluorescence techniques using labeling fluorophors have been applied to the determinations of different primary amines (34), useful amino acids (35), fatty acids (36, 37), and so on. Several extrinsic fluorophors are illustrated in Fig. 2.17, and some interesting fluorescent labels in many extrinsic fluorophors are described below.

### 7.2.1. Fluorescein and Rhodamine Isocyanates and Isothiocyanates and Dansyl Chloride (DNS–Cl)

The extrinsic fluorophors are widely used to label proteins. Fluorescein-labeled immunoglobulins are frequently used in fluorescence microscopy because of their high fluorescence quantum yields and resistance to photobleaching.

### 7.2.2. Extrinsic Fluorophors for Carboxylic Acids

The carboxylic acids of acidic components in living bodies play an important role in connection with many physiological reactions,

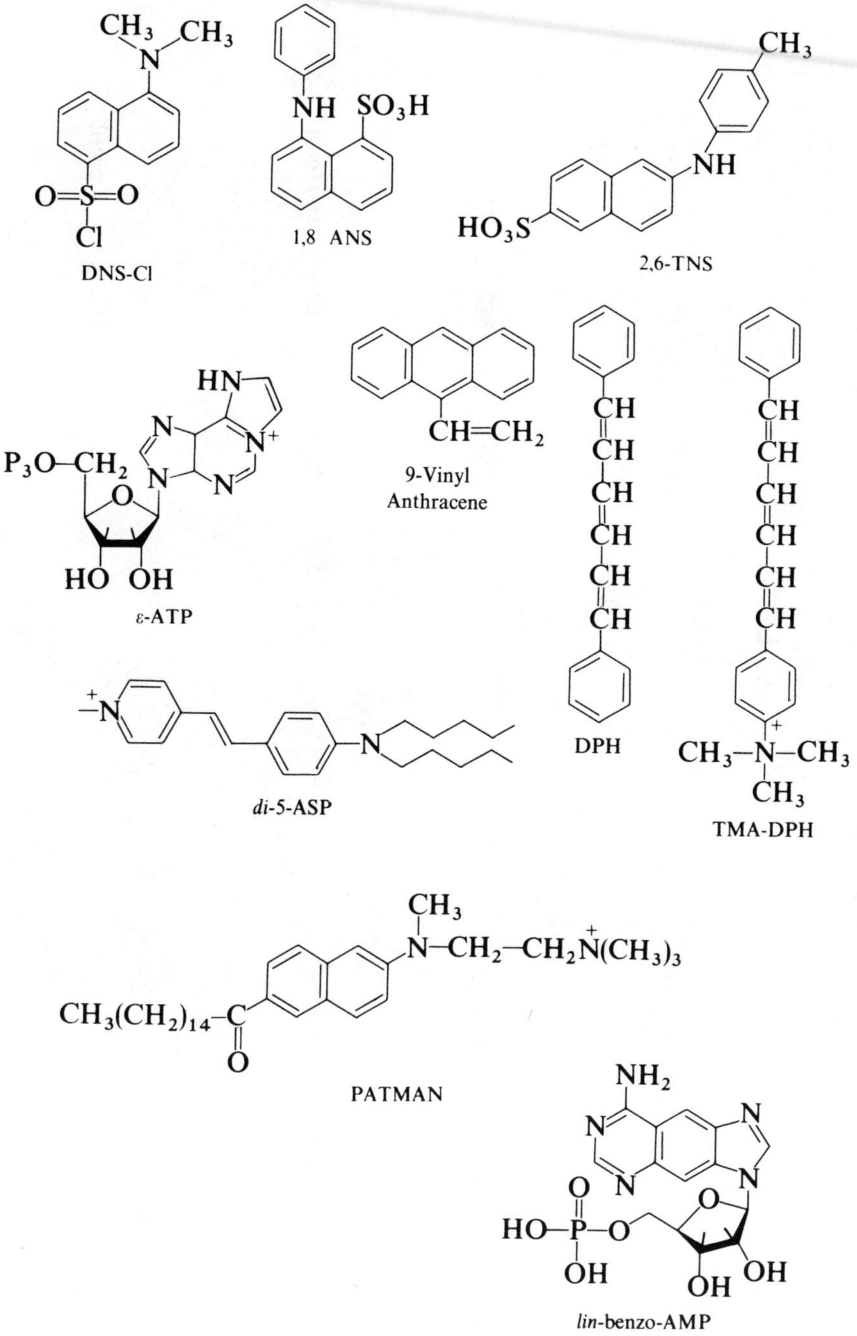

**Figure 2.17.** Several extrinsic fluorophors.

Table 2.3. Mainly Extrinsic Fluorophors for Carboxylic Acids

| Substances | Structures | $\lambda_{ex}$ | $\lambda_{em}$ | Limit of Detection |
|---|---|---|---|---|
| 4-Bromomethyl-7-methoxy-coumarin (Br-Mmc) | | 323 | 395 | 9 pmol |
| 9,10-Diamino-phenanthrene (DAPH) | | 255 | 382 | 30 fmol |
| 9-Anthryldiazomethane (ADAM) | | 365 | 412 | 100 fmol |
| 1-Bromoacetylpyrene | | 360 | 450 | 5 pmol |

| Reagent | | | |
|---|---|---|---|
| 7-Acetoxy-4-bromomethyl-coumarin (ABMC) | 346 | 450 | Several tens fmol |
| $N,N'$-Dialkyl-o-(7-methoxycoumarin-4-yl)-methylisourea | 326 | 397 | Several pmol |
| D- or L-1-Aminoethyl-4-dimethyamino-naphthalene (D- or L-DANE) | 320 | 395 | 1 pmol |

**Table 2.3.** (*Continued*)

| Substances | Structures | $\lambda_{ex}$ | $\lambda_{em}$ | Limit of Detection |
|---|---|---|---|---|
| 4-Diazomethyl-7-methoxycoumarin (DMC) | | 323 | 395 | — |
| 9-(Chloromethyl)anthracene (9-CMA) | | 365 | 412 | 2 fmol |

metabolic processes and disease states. They include prostaglandins, bile acids, α-keto acids, and so on. The mainly extrinsic fluorophores for these carboxylic acids are indicated in Table 2.3.

A fluorescence labeling reaction with 9-anthryldiazomethane (ADAM) (38) as an extrinsic fluorophor to the cabLexylic acids is shown in Eq. (2.7):

$$R-COOH + \text{ADAM (CHN}_2\text{)} \rightarrow \text{ADAM derivative (CH}_2\text{OCO-R)} + N_2 \quad (2.7)$$

## 8. BASIS OF LUMINESCENCE ANALYSIS (39)

### 8.1. The Place of Luminescence Assays in Biochemistry

Biochemists are involved in measuring a variety of substances by many different analytical techniques. Although different, these techniques share the common principle of an interface between chemistry and physics. The most commonly used such technique in biochemistry is absorptiometry, at both visible and at ultraviolet wavelengths, but emission flame photometry and radioactivity are also widely used. A novel analytical technique, so far relatively unexplored in biochemistry, is luminescence.

Analyses based on the measurement of emitted light offer several advantages over conventional techniques: high sensitivity, wide linear range, low cost per test, and relatively simple and inexpensive equipment.

Luminescence is expected to have an impact on two main areas of biochemical analysis. First, it may have an important role as a replacement for conventional colorimetric or spectrophotometric indicator reactions in assays for substrates of oxidases and dehydrogenases. In this type of assay the sensitivity of the luminescence indicator reaction may be used either to quantitate substances not easily measured by conventional techniques (e.g., prostaglandins and vitamins) or to reduce the quantities of specimen and reagent required in the initial enzymatic step, thus reducing the cost of the assay. The second major application of luminescence is the utilization of luminescent molecules as replacements for radioactive labels in immunoassay.

An important feature of luminescence as an analytical technique is the fact that not only is its usefulness confirmed in biochemistry, especially biomedical and clinical chemistry, but also its role is increasing in other pathology disciplines, such as hematology, immunology, bacteriology, and pharmacology. To enable the reader to fully appreciate this usefulness of luminescence for biochemists, we will provide a general discussion of analytical luminescence as applied to the above-mentioned fields.

## 8.2. Terminology and Definitions

Chemiluminescence may be simply defined as the chemical production of light. In the literature it is often confused with fluorescence. The difference between the two processes is the source of the energy that produces molecules in an excited state. In chemiluminescence this is the energy of a chemical reaction, and the decay from the excited state back to the ground state is accompanied by the emission of light (luminescence). In contrast, incident radiation is the source of the energy in fluorescence. Analytically, the types of luminescence that have engendered the most interest are chemiluminescence (CL) and bioluminescence (BL). The latter name is given to a special form of CL found in biological systems, in which a catalytic protein increases the efficiency of the luminescent reaction. Indeed, in certain cases the reaction is impossible without a protein component.

## 8.3. Historical Aspect

Luminescence of various types has almost certainly been known since ancient times. For example, Aristotle (384–322 B.C.) described the BL of dead fish (bacteria) and fungi in his *De Anima*. Little further progress was made in studies of BL until the nineteenth century, with the pioneering work of Dubois (40) on the BL of firefly and clam extracts. In contrast to BL, the history of CL, especially that in the aqueous phase, is remarkably short. The fortuitous discoveries of lophine (41) and pyrogallol (42) (see Fig. 2.18) were the first examples in solution. The important CL substances luminol and lucigenin were discovered in 1928 (43) and 1935 (44), respectively.

For a more complete discussion of luminescence, or indeed of any aspect of its history, the reader is referred to the remarkable book by Harvey (45).

## 8.4. Types of Luminescence

### 8.4.1. Chemiluminescence

The quantum efficiency of CL is often very low; for example, for luminol it ranges from 0.01 to 0.05 (46). However, the development of sensitive photomultipliers has made it possible to follow very weak luminescence, even when the quantum efficiency is as low as $10^{-15}$ (46).

Today many CL systems are known, but despite intensive studies the detailed mechanisms involved are often obscure. CL, although very weak, is stated to be a virtually universal property of oxidizable organic substances (47).

CL can occur in the gas, solid, or liquid phase; however, only liquid-phase CL, which appears to be the most useful for application to biochemistry, will be considered here. The mechanism of organic CL in solution involves three stages (48): (1) preliminary reactions to provide the key intermediate; (2) an excitation step, in which the chemical energy of the key intermediate is converted into electronic excitation energy; and (3) emission of light from the excited product formed in the chemical reaction. In reactions in which a fluorescent compound is added to enhance the CL emission, an efficient energy transfer occurs and the resulting CL is known as "sensitized CL."

All the CL reactions described in this section have the common feature of being hydrogen peroxide detectors.

1. *Diacylhydrazides.* The CL of a cyclic diacylhydrazide, luminol (5-amino-2,3-dihydro-1,4-phthalazinedione), was first described by Albrecht in 1928 (43). Since then, many acylhydrazides have been checked for their ability to luminescence, but strong CL is obtained only from cyclic diacylhydrazides of the same type as luminol. A shift in the position of the amino group reduces efficiency; for example, isoluminol (6-amino-2,3-dihydrophthalazine-7,4-dione) is 10% as efficient as luminol (49). Substitution in the ring structure markedly influences the luminescence. Electron-withdrawing (electrophilic) substituents in the benzene ring decrease luminescence, but electron-donating (electrophobic) substituents increase light yield, substitution at positions 5 and 8 being more effective than at 6 and 7. A complete loss of light emission occurs if the heterocyclic ring is substituted. Annelated analogs of luminol have been produced that are 300% more efficient than luminol (see Fig. 2.18) (49, 50). The efficiency, wavelength, and pH optimum of light emission in luminol depend greatly on reaction conditions.

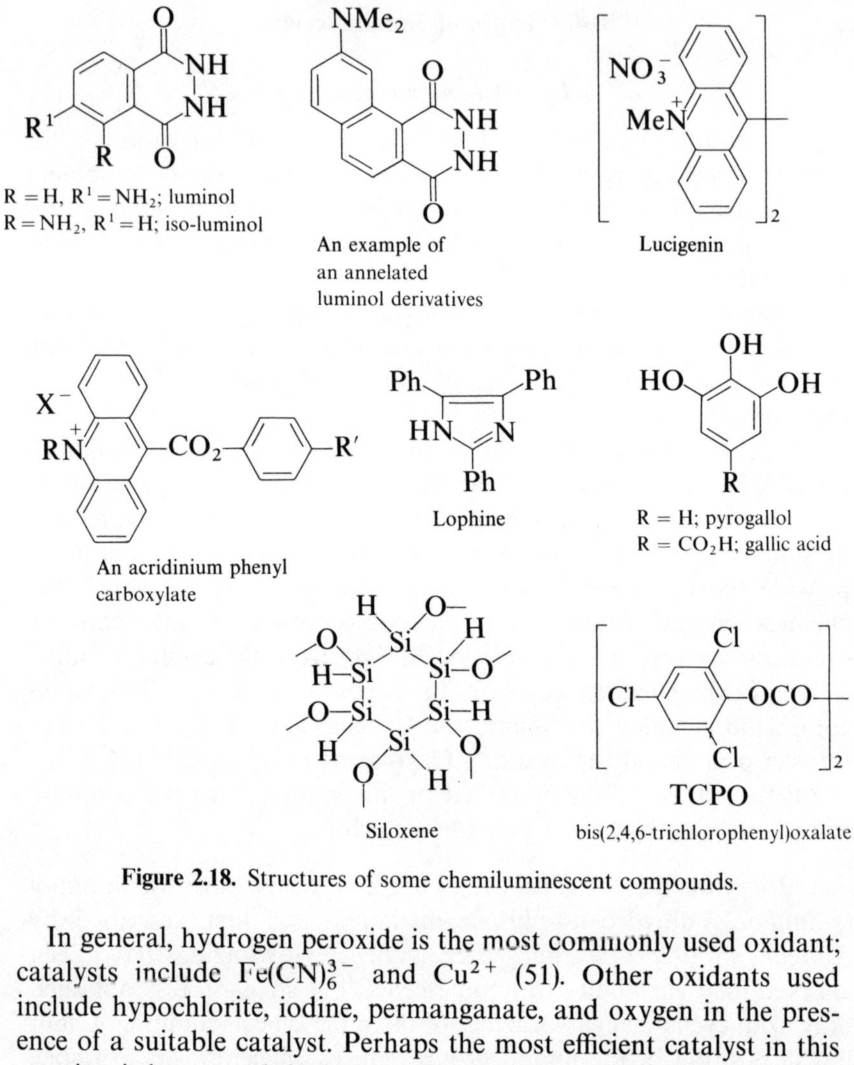

**Figure 2.18.** Structures of some chemiluminescent compounds.

In general, hydrogen peroxide is the most commonly used oxidant; catalysts include $Fe(CN)_6^{3-}$ and $Cu^{2+}$ (51). Other oxidants used include hypochlorite, iodine, permanganate, and oxygen in the presence of a suitable catalyst. Perhaps the most efficient catalyst in this reaction is heme, and the best medium would appear to be carbonate buffer (52). The optimum pH for CL varies somewhat with the catalyst and oxidant, but for most oxidizing systems it is near 11. The luminescence of luminol follows the stoichiometry in Eq. (2.8) (53), and it has been shown that the aminophthalate ion, which is formed in the singlet excited state, is the light emitter in the reaction (54):

$$\text{luminol} + H_2O_2 \xrightarrow[M^{x+}]{OH^-} \text{aminophthalate ion} + \text{light (425 nm)} \quad (2.8)$$

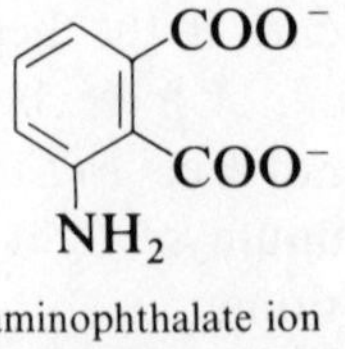

It remains in doubt how excited states are produced in the CL of the cyclic hydrazides, but electron-transfer mechanisms (52, 55), mechanisms involving the intercession of diazoquinones and dioxetanes (52), and (more recently) chemically initiated electron-exchange luminescence mechanisms have been proposed (56).

Comparison of some different oxidation systems for detecting luminol reveals that sensitivity is greatest with hydrogen peroxide and catalysts containing heme.

An interesting older observation (57) is that plant peroxidases and luminol in the presence of hydrogen peroxide will react to produce light at near-neutral pH. The detection limit for luminol with the hydrogen peroxide-microperoxidase system is 1 pmol $dm^{-3}$ (49).

2. *Acridinium salts.* Lucigenin (bis-*N*-methylacridium nitrate) is similar to luminol in that it luminescences on oxidation by peroxide in basic solution in the presence of metal-ion catalysts, the quantum efficiency of the process being about 0.01–0.02 (Eq. (2.9)):

$$\text{lucigenin} + H_2O_2 \xrightarrow[M^{x+}]{OH^-} MeN\text{\textlangle\textrangle}=O + \text{light (470 nm)} \quad (2.9)$$

*N*-methylacridone

The reaction is catalyzed by some metal ions, such as $Pb^{2+}$, that do not catalyze the luminol reaction and thus provides the basis for analytical applications not possible with luminol. Considerably less research effort has been expended on the mechanism of light production in lucigenin than in luminol. What work has been done (58, 59) is largely in relation to its enzymic reactions, especially with xanthine oxidase (EC 1.2.3.2).

McCapra et al. (60) describe novel acridinium phenylcarboxylates, which undergo CL oxidation with hydrogen peroxide. The optimum pH of oxidation of these substances depends on the nature of the substituent in the phenyl group, thus allowing an acridinium salt to be tailored to fit particular pH requirements. Their shelf life and susceptibility to interferences are claimed to be superior to those of other CL reagents.

3. *Diaryl oxalates.* Diaryl oxalates such as bis(trichlorophenyl) oxalate (TCPO) undergo a CL oxidation reaction with hydrogen peroxide (61–64), via a peroxyoxalate intermediate (Eq. (2.10)):

$$\text{TCPO} + \text{H}_2\text{O}_2 \longrightarrow \text{Cl-C}_6\text{H}_2\text{Cl}_2\text{-OCOCOOOH} + \text{Cl-C}_6\text{H}_2\text{Cl}_2\text{-OH}$$

[1]           [2]

$$[1] \longrightarrow [3] + [2]$$

[3] peroxyoxalate intermediate

$$3 + \text{fluorescer (F)} \longrightarrow F^* + 2\,CO_2$$
$$F^* \longrightarrow F + \text{light} \qquad (2.10)$$

Efficient CL occurs in ester solutions, and the intensity is increased by organic bases and inhibited by organic acids (65). A problem with diaryl oxalates is their susceptibilities to hydrolysis and the subsequent effect on light-emission characteristics. TCPO is one of the more stable diaryl oxalates, and no hydrolysis-related problems have been encountered in its analytical applications. It has been used, in an ethyl acetate/methanol/aqueous buffer (pH range 4–10) system containing triethylamine, for the analysis of peroxide. A fluorescent molecule, perylene, has been used as the light emitter in this reaction (sensitized CL) because of its stability and its favorable efficiency and wavelength range of emission (61). The detection limit is $7 \times 10^{-8}$ mol dm$^{-3}$, and the linear response, range extends up to $10^{-3}$ mol dm$^{-3}$ of hydrogen peroxide. Although the bis(trichlorophenyl) oxalate system is not as sensitive as luminol for the detection of peroxide, it does have the advantages of a lower background CL and less sensitivity to sources of interference such as uric acid.

4. *Other CL systems.* Many other liquid-phase CL systems than luminol and lucigenin have been described. For example, lophine is oxidized under basic conditions to give a yellow CL (41), and siloxine (66) is oxidized in acidic media to produce a yellow-red CL, depending on the oxidant and the degree of oxidation. Polyhydric phenols such as pyrogallol (67) and gallic acid (67) also undergo luminescent oxidations. Many other CL reactions are known in solution, but their usefulness in analysis has not been explored.

### 8.4.2. *Bioluminescence*

Numerous assays for substances of biochemical interest are based on reactions involving cofactors such as NAD$^+$/NADH and ATP. The

BL reactions described in the following sections offer sensitive alternative to the conventional spectrophotometric or colorimetric assays for such cofactors.

Although terrestrial luminescent organisms such as the firefly and glowworm are the best-known BL organisms, most of the other examples are sea-living organisms, ranging in complexity from microscopic bacteria and plankton to many species of fish. So widespread a phenomenon is BL that two thirds of the organisms in the upper 2,000 m of oceanic waters are BL (68). To date, fewer than 1% of the known luminous biological species have been studied in great detail (69), but sufficient data have been accumulated to allow the various types of BL to be classified mechanistically (70) as follows:

1. Pyridine-nucleotide linked systems.
2. Adenine-nucleotide linked systems.

### Table 2.4. Classification of BL Reactions[a]

1. Pyridine-nucleotide linked
   e.g., *Vibrio fischeri*, *Beneckea harveyi* (marine bacteria)

   $$NADH + FMN \xrightarrow{reductase} FMNH_2 + NAD^+$$

   $$FMNH_2 + RCHO + O_2 \xrightarrow{luciferase} FMN + RCO_2H + light$$

2. Adenine-nucleotide linked
   e.g., *Photinius* (firefly)

   $$Luciferin + ATP + O_2 \xrightarrow[Mg^{2+}]{luciferase} ADP + CO_2 + light$$

3. Enzyme-substrate systems
   e.g., *Pholas* (clam)

   $$Luciferin + O_2 \xrightarrow{luciferase} light$$

5. Activation of "precharged" systems
   e.g., *Aequorea* (a coelenterate)

   $$Precharged\ protein \xrightarrow{Ca^{2+}} blue\text{-}fluorescent\ protein + light$$

---

[a] Type 4 (peroxidation systems) have not as yet found application in clinical chemistry.

3. Enzyme-substrate systems.
4. Peroxidation systems.
5. "Precharged" systems

Table 2.4 gives examples of the most important reaction types in terms of applications in biochemistry.

**8.4.2.1. Firefly.** Firefly luminescence undoubtedly has been the most extensively studied BL system. The light-producing reaction requires the enzyme luciferase, luciferin, $Mg^{2+}$, ATP, and molecular oxygen. Numerous reviews deal with the reaction (70–75), and the mechanism now proposed is shown in Fig. 2.19 (76–79). The initial reaction is the rapid conversion, in the presence of $Mg^{2+}$ and ATP, of luciferin to luciferyl adenylate, which, in the presence of luciferase, combines with molecular oxygen to give an oxyluciferyl adenylate-enzyme complex in the excited state. After emission, the ground-state complex disassociates to form enzyme, ATP, oxyluciferin, and carbon dioxide, the last being derived from the carboxyl group of luciferin. The reaction is best carried out at 25°C in glycine buffer, pH 7.8 (77). The color of the light emitted differs for different species of firefly, and this is thought to be due to interspecies differences in the structure of luciferase, the structure of luciferin being identical for all species (80). Intensity and wavelength of maximal emission are also altered by changes in pH, ionic strength, temperature, and the presence of chlorides of $Zn^{2+}$ or $Cd^{2+}$ (81).

$$LH_2 + E + ATP \xrightarrow{Mg^{2+}} E:LH_2:AMP + PP$$

$$E:LH_2:AMP + O_2 \longrightarrow E + L{=}O + CO_2 + AMP + \text{light}$$

$hv_{max} = 562$ nm, $\phi BL \cong 1$

$\phi BL$; quantum yield, $LH_2$; luciferin, E: luciferase
$E:LH_2:AMP$; luciferyl adenylate, PP; pyrophosphate
$L{=}O$; oxyluciferin

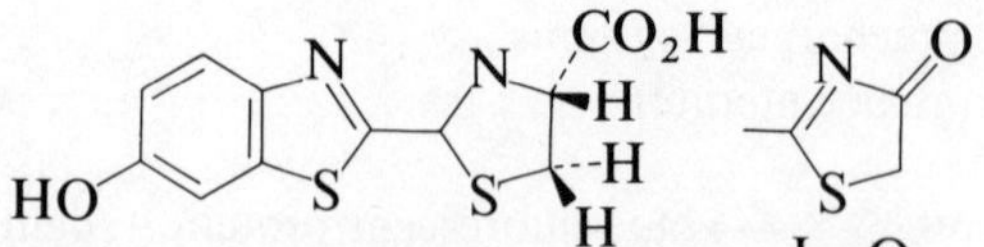

**Figure 2.19.** Mechanism of firefly luminescence.

***8.4.2.1.1. Luciferase.*** Firefly luciferase is a dimer, each subunit having a relative molecular mass of 50,000. Only one of the subunits exhibits enzyme activity in the BL reaction. Studies of its amino acid composition have shown luciferase to be one of the most hydrophobic enzymes known (82) and this may explain its poor solubility. The activity of firefly luciferase preparations depends greatly on the mode of preparation. Procedures involving precipitation with $(NH_4)_2SO_4$ of firefly-lantern extracts give the best preparations, which may be stored at $-20°C$ (83–85). Enzyme activity has also been stabilized by bovine serum albumin, sucrose, ascorbic acid, or thioethanol (86, 87).

***8.4.2.1.2. Specificity.*** The reaction is not specific for ATP; other nucleotides such as cytidine-5'-triphosphate, inosine-5'-triphosphate, and iso-ATP (88) can stimulate light emission. ADP, uridine triphosphate, and guanine triphosphate have also been shown by some workers to stimulate light emission, but these results are controversial (89–91). Manganese cations may replace $Mg^{2+}$ in the reaction (92).

***8.4.2.1.3. Inhibitors.*** Certain anions inhibit the reaction: $SCN^- < I^- < NO_3^- < BR^- < Cl^-$ (93). Anesthetics such as procaine and lidocaine are also inhibitors, and this fact has formed the basis of assays (94).

**8.4.2.2. Marine Bacteria.** Most studies of luminescent marine bacteria have centered on two types, *Benecka harveyi* (*Photobacterium fischeri* strain MAV) and *Vibrio fischeri*. Characteristics of the reaction such as kinetics, absolute quantum yields, and emission spectra vary with the type of bacterium from which the luciferase is obtained. The BL quantum yield per enzyme turned over, or product formed, is in the range 0.1–0.2 (95). Generally the emission spectra are characterized by a broad emission, with a maximum near 478–505 nm. Recently, however, a yellow-emitting strain of *P. fischeri* has been described (96).

*In vitro*, the components required for luminescence are $FMNH_2$ generated from FMN by the oxidation of NADH or NADPH with the aid of FMN reductase, a long-chain aliphatic aldehyde, oxygen, and bacterial luciferase. The proposed sequence and the intermediate and alternative pathways are presented in Fig. 2.20. The total light produced in the reaction is proportional to the amount of each of the substrates ($O_2$, $FMNH_2$, and RCHO) when they are present in limiting quantities. The same can be said of the luciferase because

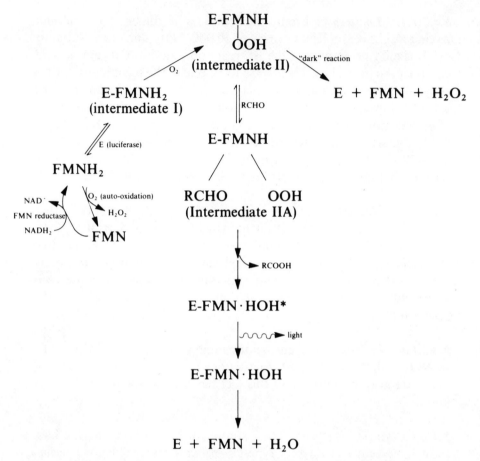

**Figure 2.20.** The reaction sequence and intermediates postulated for the luciferase-mediated bioluminescent oxidation of $NADH_2$ including nonenzymatic and dark pathways.

excess $FMNH_2$ is auto-oxidized so rapidly, as compared with the rate of light emission, that $FMNH_2$ has been reconverted to FMN by the time the luciferase finishes its catalytic cycle. Thus, like the other reactants, the luciferase acts only once in the *in vitro* reaction. It is, however, capable of turnover on repeated addition of $FMNH_2$. Although the role of the aldehyde was not understood for a long time, it is now established that the aldehyde is oxidized to the corresponding carboxylic acid (95). Various induced mutants of *Beneckea harveyi* are known (95). These show altered BL properties, such as color of light emission, temperature sensitivity, and luciferase turnover rate. Recently, a dim mutant requiring myristic acid (97) and a nonluminous mutant that produces light only in the presence of cAMP have been described (98), but their analytical potential has not yet been fully explored.

***8.4.2.2.1. Luciferase.*** Although luciferases from different strains have similar requirements and appear to involve similar reaction pathways, the occurrence of two distinctly different luciferases has been reported (99). The luciferases from *Vibrio fischeri* and *Beneckea harveyi* have been shown to be heteropolymers composed of two nonidentical subunits, and contain no metals or other cofactors (100). The $\alpha$- and $\beta$-subunits of luciferase from these bacteria have relative molecular masses of 41,000 and 38,000, and 42,000 and 37,000, respectively. No active hybrids are formed between subunits of the two luciferases, and tryptic peptides of the two luciferases also differ (101).

The role of each subunit in the BL reaction appears to be quite different, and it has been shown that the active site is located specifically on the $\alpha$-subunit. The role of the $\beta$-subunit remains to be elucidated, although it is essential for BL activity (102).

Luciferase contains between 12 and 15 sulfhydryl groups, and luciferase from *Beneckea harveyi* has a reactive sulfhydryl group on the $\alpha$-subunit, at or near the active center (103). Treatment of bacterial luciferase with 8 mol dm$^{-3}$ urea or 5 mol dm$^{-3}$ guanidine hydrochloride leads to separation of the subunits and rapid loss of activity. The subunits can be recombined and the activity recovered, but this recovery is absolutely dependent on keeping the sulfhydryl groups reduced (104).

***8.4.2.2.2. pH Profile.*** The two luciferases show different pH profiles. *Vibrio fischeri* luciferase shows a broad profile with nearly optimal activity over the pH range 6.4–7.2. *Beneckea harveyi* luciferase shows a quite different response, having its optimum (measured by use of the initial maximal light intensity) in the region of pH 5.6–6.8, declining sharply at higher values (99).

***8.4.2.2.3. Stability.*** Thermal-stability studies in which the enzyme is incubated at various temperatures for 5 min and the residual enzymic activity then determined, have shown that luciferase activity begins to lessen between 25 and 35°C (105), presumably because of heat denaturation of the enzyme. At 40°C this process takes just a few minutes. Heat-denatured luciferase may be reactivated by dissolving it in 8 mol dm$^{-3}$ of urea and diluting (104). Luciferase is inactivated by proteases (106), but this effect is reduced dramatically on binding of FMN, phosphate, or sulfate anions (107).

*8.4.2.2.4. Specificity.* Luciferase activity is highly specific for $FMNH_2$, but the enzyme also shows weak activity toward other flavins and flavin analogs (108, 109). Only aliphatic aldehydes with chain length of eight or more carbon atoms are effective in the luminescent reaction.

*8.4.2.2.5. Inhibitors.* Luciferase is particularly sensitive to thiol reagents such as p-chloromercuribenzoic acid and to reagents that react with lysyl, cysteinyl, and histidinyl residues (95). Volatile anesthetics (110), riboflavin, cyanide, copper, iron, and other heavy metals also inhibit the enzyme (111).

**8.4.2.3. FMN Reductase.** Flavin mononucleotide reductase (flavin reductase, NAD(P)H:dehydrogenase, or NAD(P)H-FMN oxidoreductase) appears to associate *in vitro* with bacterial luciferase (112). The reductase is also postulated to supply reduced FMN *in vivo* as a substrate for the BL reaction. The enzyme is present not only in luminescent bacteria but also in aerobic and anaerobic nonluminescent bacteria such as *Escherichia coli* (113).

There is still some controversy over the types of FMN reductase present in the cell. Separation of two reductases, a free form and one associated with luciferase, has been reported. Other workers maintain, however, that only a single, free reductase exists (112).

Distinct FMN reductases specific for NADH and NADPH have been isolated from extracts of *Beneckea harveyi* by affinity chromatography (114). The NADH- and NADPH-specific FMN reductases have relative molecular masses ($M_r$) of 30,000 and 40,000, and $K_m$'s of $4.75 \times 10^{-5}$ mol dm$^{-3}$ of NADH and $4.0 \times 10^{-5}$ mol dm$^{-3}$ of NADPH.

**8.4.2.4. Aequorea.** The BL system of the jellyfish *Aequorea* consists of protein-chromophore complexes (termed *photoproteins*), which react with $Ca^{2+}$ to produce a bluish luminescence ($\lambda_{em} = 469$ nm) independent of dissolved oxygen. The photoprotein consists of a species-specific protein in close association with a chromophore component (luciferin), which is oxidized by the protein in the presence of $Ca^{2+}$ to oxyluciferin, with production of light. The spent photoprotein is termed *blue-fluorescent* (*BFP*) (115, 116).

*8.4.2.4.1. Specificity.* Metal ions other than $Ca^{2+}$ trigger the BL of the *Aequorea* system, notably $Sr^{2+}$, $Ba^{2+}$, and all of the lanthanides (116).

**8.4.2.5. Other BL Systems.** Many other BL organisms are known, for example, fungi (*Collybia velutipes, Armillaria mellea*), sea pansy (*Renilla reniformis*), marine worm (*Chaetopterus*), earthworm (*Otochaetus, Diplocardia*), protozoa (*Gonyaulax polyedra*), fish (*Apogon, Photoblepheron*), shrimp (*Heterocarpus*), beetle (*Pyrophorus*), but generally their requirements for BL and mechanisms are poorly characterized (117). We anticipate that analytically useful BL reactions may be discovered among such organisms.

The extreme sensitivity of luminescence assays and their applicability to the important analytical intermediates, hydrogen peroxide, ATP, and NAD(H) have led to the exploitation of CL and BL as analytical tools in biochemistry. The assays based on CL, firefly BL, and bacterial BL are assembled in Tables 2.5, 2.6, and 2.7.

Table 2.5. Methods Based on CL ($H_2O_2$ Detection)

| Analyte | Sensitivity or Range | Reference |
|---|---|---|
| Vitamin $B_{12}$ | $2 \times 10^{-9}$ mol dm$^{-3}$ | 118 |
| Catalase | $10^{-4}$ μg | 119 |
| Chromium(III) | — | 120, 121 |
| Cytochrome $c$ | $10^{-1}$ μg | 119 |
| Ferritin | ~1 μg | 119 |
| Glucose | $2 \times 10^{-8}$ mol dm$^{-3}$ | 61, 122–125 |
| Hematin | $10^{-5}$ μg | 119 |
| Hemoglobin | $10^{-4}$ μg | 119 |
| Hydrogen peroxide | $10^{-9}$ mol dm$^{-3}$ | 51 |
| Hypoxanthine | | 126, 127 |
| Lactate dehydrogenase | — | 64 |
| Myoglobin | $10^{-4}$ μg | 119 |
| NADH | $2 \times 10^{-7}$–$10^{-4}$ mol dm$^{-3}$ | 64 |
| Uric acid | $10^{-7}$–$10^{-5}$ mol dm$^{-3}$ | 128 |

Table 2.6. Assays Based on Firefly BL (ATP Measurement)

| Analyte | Reference |
|---|---|
| *Substances* | |
| ADP | 85, 129, 130 |
| Adenosine phosphosulfate | 131 |
| AMP | 85 |
| cAMP | 86 |
| ATP | 129, 130, 132 ~ 135 |
| Adenosine tetraphosphate | 136 |
| Coenzyme A | 137 |
| Creatine | 85 |
| Creatine phosphate | 85, 135 |
| Cytidine triphosphate | 85 |
| Glucose | 85 |
| Glycerol | 138 |
| Guanosine triphosphate | 85 |
| Phosphoenolpyruvic acid | 85 |
| Pyrophosphate | 139 |
| Triglycerides | 138 |
| Uridine triphosphate | 85 |
| *Enzymes* | |
| Adenosine 3',5'-monophosphate phosphodiesterase | 86 |
| Apyrase | 85 |
| ATPase | 85 |
| ATP sulfurylase | 131 |
| Creatine kinase | 85, 140 ~ 143 |
| Creatine kinase (B subunit) | 140 |
| Guanosine 3',5'-monophosphate phosphodiesterase | 144 |
| Hexokinase | 85 |
| Myokinase | 85 |
| Nucleotide phosphokinases | 85 |
| Pyruvate kinase | 85 |

## Table 2.7. Assays Based on Bacterial BL [NAD(P)H] Measurement

| Analyte | Sensitivity or Range | Reference |
|---|---|---|
| *Substrates* | | |
| Acetylpyridine-NADH | 0.1–1.4 pmol | 145 |
| Aldehydes | | 85 |
| Ammonia | | 146 |
| Ethyl alcohol | 0.0004–0.015% | 147, 148 |
| FAD | | 149 |
| Fatty acids | | 97 |
| FMN | | 85, 134, 149, 150 |
| $FMNH_2$ | | 85, 150 |
| Glucose | 150–1500 pmol | 145, 147 |
| Glucose 6-phosphate | | 151 |
| Glucose 1-phosphate | | 151 |
| L-Glycerol 1-phosphate | | 152 |
| Glycogen | | 151 |
| 3-Hydroxybutyrate | | 152 |
| Malate | | 145, 152 |
| $NAD^+$ | | 85, 134, 145, 150, 153 |
| NADH | $10^{-14}$–$10^{-7}$ mol | 134, 145 |
| $NADP^+$ | | 134, 154 |
| NADPH | 10 pmol–200 nmol | 134, 154, 155 |
| Oxaloacetate | | 139, 150 |
| Oxygen | | 85 |
| Pyruvate | | 151 |
| *Enzymes* | | |
| Alcohol dehydrogenase | 0.015–3 pmol | 147 |
| ATP:NMN adenylyltransferase | | 153 |
| Glucose-6-phosphate dehydrogenase | 0.0015–0.1 pmol | 147 |
| Hexokinase | 0.1–2.0 pmol | 147 |
| D-3-Hydroxybutyrate dehydrogenase | | 151 |
| Isocitrate dehydrogenase | | 156 |
| Lactate dehydrogenase | 0.003–0.7 pmol | 147 |
| Malate dehydrogenase | 0.007–0.7 pmol | 147 |
| Proteolytic enzymes | | 106 |
| Trypsin | 20 ng | 106 |

## REFERENCES

1. A. Einstein, *Ann. Phys.*, **17**, 132 (1905).
2. J. Frank, *Trans. Faraday Soc.*, **21**, 536 (1926); Condon, E. U., *Phys. Res.*, **28**, 1182 (1926); **32**, 858 (1928).
3. A. Jablonski, *Z. Phys.*, **94**, 38 (1935).
4. R. S. Becker, *Theory and Interpretation of Fluorescence and Phosphorescence*, Wiley Interscience, New York, 1969.
5. N. Mataga and T. Kubota, *Molecular Interactions and Electric Spectra*, M. Dekker, New York, 1970.
6. I. B. Berlman, *Energy Transfer Parameters of Aromatic Compounds*, Academic Press, New York, 1973.
7. M. Zander, *Fluorimetrie*, Springer-Verlag, Berlin, Heidelberg, New York, 1981.
8. G. Schwedt, *Fluorimetrische Analyse, Methoden und Anwendungen*, Verlage Chemie, Weinheim, Deerfield Beach, Florida, Basel, 1981.
9. G. W. Robinson and R. P. Forsch, *J. Chem. Phys.*, **37**, 1962 (1962); **38**, 1187 (1963).
10. M. Couterman, *J. Chem. Phys.*, **36**, 2846 (1962).
11. G. N. Lewis and M. Kasha, *J. Am. Chem. Soc.*, **66**, 2100 (1944).
12. G. G. Stokes, *Phil. Trans. R. Soc. London*, **142**, 463 (1852).
13. M. Kasha, *Disc. Faraday Soc.*, **9**, 14 (1950).
14. B. L. Van Duuren, *Chem. Rev.*, **63**, 325 (1963).
15. E. J. Bowen and J. Sahn, *J. Phys. Chem.*, **63**, 4 (1959).
16. W. R. Ware and B. A. Baldwin, *J. Chem. Phys.*, **43**, 1194 (1965).
17. S. G. Schulman, *Rev. Anal. Chem.*, **1**, 85 (1971).
18. C. A. Parker and W. T. Rees, *Analyst*, **87**, 83 (1962).
19. T. Förster, *Naturwissenschaften.*, **33**, 220 (1946).
20. W. West, in *Chemical Applications of Spectroscopy* (Techniques in Organic Chemistry, Vol. 9), Wiley Interscience, New York, 1956.
21. D. M. Hercules, *Fluorescence and Phosphorescence Analysis—Principles and Applications*, Wiley Interscience, New York, 1966.
22. E. L. Wehry and L. B. Rogers, in ref., 21, pp. 81–99.
23. G. G. Guilbault, *Practical Fluorescence—Theory, Method and Techniques*, M. Dekker, New York, 1973.
24. H. Gotô, *J. Chem. Soc. Jpn.* (*Nippon Kagaku Zashi*), **56**, 199 (1938); *Sci. Rept. Tohoku Univ.*, 1 Ser. **28**, 458 (1940).
25. M. Kasha, *Radiation Res. Supl.*, **2**, 243 (1960).
26. A. Weller and W. Uraban, *Angew. Chem.*, **66**, 336 (1954).
27. D. M. Hercules and B. Rogers, *Spectrochim. Acta*, **15**, 393 (1959).
28. Z. Yoshida and R. Oda, *Mem. Faculty Eng. Kyoto Univ.*, **13**, 108 (1951).

29. J. Marks, *The Vitamins in Health and Disease*, J. and A. Churchill Ltd., London, 1968, p. 85.
30. J. Koziol, *Photochem. Photobiol.*, **5**, 41 (1966).
31. C. Y. W. Ang and F. A. Mosely, *J. Agr. Food Chem.*, **28**, 483 (1980).
32. J. R. Bensen and P. E. Hare, *Proc. Natl. Acad. Sci. U.S.A.*, **72**, 619 (1975).
33. K. Samejima, W. Dairman, and S. Udenfriend, *Anal. Biochem.*, **42**, 222 (1972); *Science*, **178**, 871 (1972).
34. S. Stein, P. Böhlen, and J. Stone, et al., *Arch. Biochem. Biophys.*, **155**, 203 (1973).
35. S. DeBernardo, M. Weigele, and V. Toome, et al., *Arch. Biochem. Biophys.*, **163**, 390 (1974).
36. N. Ichinose, K. Nakamura, and C. Shimizu, et al., *J. Chromatogr.*, **295**, 463 (1984).
37. N. Ichinose, K. Nakamura, and C. Shimizu, et al., *Jpn. Soc. Anal. Chem. (Bunseki Kagaku)*, **33**, E271 (1984).
38. N. Nimura and T. Kinoshita, *Anal. Lett.*, **13**, 191 (1980); *Shimazu Science and Instruments News*, **24**, 9 (1983).
39. T. P. Whitehead, L. J. Kricka, T. J. N. Carter, and G. H. G. Thorpe, *Clin. Chem.*, **25**, 1531 (1979).
40. R. Dubois, *C.R. Soc. Biol.* (Ser. 8), **2**, 559 (1885).
41. B. Radzizewski, *Chem. Ber.*, **10**, 70 (1877).
42. J. M. Eder, *Photogr. Mitth.*, **24**, 74 (1887).
43. H. O. Albrecht, *Z. Phys. Chem.*, **136**, 321 (1928).
44. K. Gleu and W. Petsch, *Angew. Chem.*, **48**, 57 (1935).
45. E. N. Harvey, *A History of Luminescence from the Earliest Times until 1900*, The American Philosophical Society, Philadelphia, PA, 1957.
46. U. Isacsson and G. Wettermark, *Anal. Chim. Acta*, **68**, 339 (1974).
47. G. D. Mendenhall, *Angew. Chem.*, **16**, 225 (1977).
48. C. T. Peng, in *Liquid Scintillation, Science and Technology*, edited by A. S. Noujaim, C. Ediss, and L. L. Weibe, Academic Press, New York, 1976, pp. 313–330.
49. H. R. Schroeder and F. M. Yeager, *Anal. Chem.*, **50**, 1114 (1978).
50. F. McCapra, *Q. Rev. (London)*, **20**, 485 (1966).
51. W. R. Seitz and M. P. Neary, *Anal. Chem.*, **46**, 188A (1974).
52. E. H. White and R. B. Brundrett, in *Chemiluminescence and Bioluminescence*, edited by M. J. Cormie, D. M. Hercules, and J. Lee, Plenum Press, New York, 1973, pp. 231–244.
53. E. H. White, O. C. Zafiriou, H. M. Kagi, and J. H. M. Hill, *J. Am. Chem. Soc.*, **86**, 940 (1964).
54. E. H. White and M. M. Bursey, *J. Am. Chem. Soc.*, **86**, 941 (1964).
55. E. Rapaporte, M. W. Cass, and E. H. White, *J. Am. Chem. Soc.*, **94**, 3153 (1972).

56. J-Y. Koo and G. B. Schuster, *J. Am. Chem. Soc.*, **100**, 4496 (1978).
57. E. N. Harvey, *Living Light*, Princeton University Press, Princeton, 1940, pp. 118–148.
58. J. R. Totter, V. J. Medina, and J. L. Scoseria, *J. Biol. Chem.*, **235**, 238 (1960).
59. J. R. Totter, in *Bioluminescence in Progress*, edited by F. H. Johnson and Y. Hanida, Princeton University Press, Princeton, 1966, pp. 23–33.
60. F. McCapra, D. E. Tutt, and R. M. Topping, British Patent 1,461,877 (1977).
61. D. C. Williams, G. G. Huff, and W. R. Seitz, *Anal. Chem.*, **48**, 1003 (1976).
62. P. A. Sherman, J. Holtzbecker, and D. E. Ryan, *Anal. Chim. Acta*, **97**, 21 (1978).
63. W. R. Seitz and M. P. Neary, *Methods Biochem. Anal.*, **23**, 161 (1976).
64. D. C. Williams and W. R. Seitz, *Anal. Chem.*, **48**, 1478 (1976).
65. K. Puget, A. M. Michelson, and S. Avrameas, *Anal. Biochem.*, **79**, 447 (1977).
66. L. Erdey, in *Indicators*, edited by E. Bishop, Pergamon Press, Oxford, U.K., 1972, pp. 709–732.
67. D. Slawinska and J. Slawinska, *Anal. Chem.*, **47**, 2101 (1975).
68. F. McCapra, *Acc. Chem. Res.*, **9**, 201 (1976).
69. F. McCapra, *Endeavour*, **32**, 139 (1973).
70. M. J. Cormier and J. R. Totter, *Annu. Rev. Biochem.*, **33**, 431 (1964).
71. M. J. Cormier, D. M. Hercules, and J. Lee, *Chemiluminescence and Bioluminescence*, Plenum Press, New York, 1973.
72. M. De Luca, *Adv. Enzymol. Relat. Areas Mol. Biol.*, **44**, 37 (1976).
73. H. H. Seliger, *Photochem. Photobiol.*, **21**, 355 (1975).
74. J. Lee, *Photochem. Photobiol.*, **20**, 535 (1974).
75. J. W. Hastings and T. Wilson, *Photochem. Photobiol.*, **23**, 461 (1976).
76. M. J. Cormier, J. E. Wampler, and K. Hori, *Progr. Chem. Org. Nat. Prods.*, **30**, 1 (1973).
77. H. H. Seliger and R. A. Morton, *Photophysiology*, **4**, 253 (1968).
78. E. H. White, J. D. Miano, and M. Umbreit, *J. Am. Chem. Soc.*, **97**, 198 (1975).
79. J. Y. Koo, S. P. Schmidt, and G. B. Schuster, *Proc. Natl. Acad. Sci. USA*, **75**, 30 (1978).
80. Y. Kishi, S. Matsuura, and S. Inoue, et al., *Tetrahedron Lett.*, **23**, 2847 (1968).
81. H. H. Seliger and R. A. Morton, *Photophysiology*, **4**, 296 (1968).
82. M. DeLuca, G. W. Wirtz, and W. D. McElroy, *Biochemistry*, **3**, 935 (1964).
83. B. L. Strehler and W. D. McElroy, *Methods Enzymol.*, **3**, 871 (1957).

84. B. L. Strehler, in *Methods of Enzymatic Analysis*, edited by H. U. Bergmeyer, Academic Press, New York, 1963.
85. B. L. Strehler, *Methods Biochem. Anal.*, **16**, 99 (1968).
86. R. A. Jonson, J. G. Harman, A. E. Broadus, and E. W. Sutherland, *Anal. Biochem.*, **35**, 91 (1970).
87. L. H. Kilbert, M. S. Schiff, and P. P. Foa, *Horm. Metab. Res.*, **4**, 242 (1972).
88. N. J. Leonard and R. A. Laursen, *Biochemistry*, **4**, 365 (1965).
89. J. B. St John, *Anal. Biochem.*, **37**, 409 (1970).
90. M. S. P. Manandhar and K. van Dyke, *Microchem. J.*, **19**, 42 (1974).
91. J. A. Davison and G. H. Fynn, *Anal. Biochem.*, **58**, 632 (1974).
92. W. D. McElroy, *Methods Enzymol.*, **2**, 851 (1955).
93. J. L. Denburg and W. D. McElroy, *Arch. Biochem. Biophys.*, **141**, 668 (1970).
94. I. Ueda, H. Kamaya and H. Eyring, *Proc. Natl. Acad. Sci. USA*, **73**, 481 (1976).
95. J. W. Hasting and K. H. Nealson, *Annu. Rev. Microbiol.*, **31**, 549 (1977).
96. E. G. Ruby and K. H. Nealson, *Science*, **196**, 432 (1977).
97. S. Ulitzur and J. W. Hastings, *Proc. Natl. Acad. Sci. USA*, **75**, 266 (1978).
98. S. Ulitzur and J. Yashphe, *Biochim. Biophys. Acta*, **404**, 321 (1975).
99. J. W. Hastings, K. Weber, and J. Friedland, et al., *Biochemistry*, **8**, 4681 (1969).
100. M. J. Cormier, J. Lee, and J. E. Wampler, *Annu. Rev. Biochem.*, **44**, 255 (1975).
101. A. Gunsalus–Miguel, E. A. Meighen, K. Nealson, and J. W. Hastings, *J. Biol. Chem.*, **247**, 398 (1972).
102. J. Cousineau and E. A. Meighen, *Biochemistry*, **15**, 4992 (1976).
103. M. Z. Nicoli, E. A. Meighen, and J. W. Hastings, *J. Biol. Chem.*, **249**, 2385 (1974).
104. J. Friedland and J. W. Hastings, *Biochemistry*, **6**, 2893 (1967).
105. E. Gerlo and J. Charlier, *Eur. J. Biochem.*, **57**, 461 (1975).
106. D. Njus, T. O. Baldwin, and J. W. Hastings, *Anal. Biochem.*, **61**, 280 (1974).
107. M. Z. Nicoli, T. O. Baldwin, J. E. Becvar, and J. W. Hastings, in *Flavins and Flavoproteins*, edited by T. P. Singer, Elsevier, Amsterdam, 1976, pp. 53–61.
108. E. A. Meighen and R. E. MacKenzie, *Biochemistry*, **12**, 1482 (1973).
109. G. W. Mitchell and J. W. Hastings, *J. Biol. Chem.*, **244**, 2572 (1969).
110. G. Adey, B. Wardley–Smith, and D. White, *Life Sci.*, **17**, 1849 (1976).
111. A. A. Green and W. D. McElroy, *Methods Enzymol.*, **2**, 857 (1955).
112. W. Duane and J. W. Hastings, *Mol. Cell. Biochem.*, **6**, 53 (1975).

113. K. Puget and A. M. Michelson, *Biochemie*, **54**, 1197 (1972).
114. E. Jablonski and M. DeLuca, *Biochemistry*, **16**, 2932 (1977).
115. J. R. Blinks, F. G. Prendergast, and D. G. Allen, *Pharmacol. Rev.*, **28**, 1 (1976).
116. J. R. Blinks, *Photochem. Photobiol.*, **27**, 423 (1978).
117. W. R. Seitz and M. P. Neary, *Contemp. Topics Anal. Clin. Chem.*, **1**, 49 (1977).
118. T. L. Sheehan and D. M. Hercules, *Anal. Chem.*, **49**, 446 (1977).
119. H. A. Neufeld, C. J. Conklin, and R. D. Towner, *Anal. Biochem.*, **12**, 303 (1965).
120. S. D. Hoyt and J. D. Ingle, Jr., *Anal. Chim. Acta*, **87**, 163 (1976).
121. R. T. Li and D. M. Hercules, *Anal. Chem.*, **46**, 916 (1974).
122. D. T. Bostick and D. M. Hercules, *Anal. Chem.*, **47**, 447 (1975).
123. J. P. Auses, S. L. Cook, and S. T. Maloy, *Anal. Chem.*, **47**, 244 (1975).
124. D. C. Williams, G. F. Huff, and W. R. Seitz, *Anal. Chem.*, **48**, 1003 (1976).
125. D. T. Bostick and D. M. Hercules, *Anal. Lett.*, **7**, 347 (1974).
126. J. M. Oyamburo, C. E. Prego, E. Prodanov, and H. Soto, *Biochim. Biophys. Acta*, **205**, 190 (1970).
127. J. R. Totter, V. J. Medina, and J. L. Scoseria, *J. Biol. Chem.*, **235**, 238 (1960).
128. F. Gorus and E. Schram, *Arch. Int. Physiol. Biochem.*, **85**, 981 (1977).
129. H. Holmsen, E. Storm, and H. J. Day, *Anal. Biochem.*, **46**, 489 (1972).
130. H. Hammar, *Acta Dermatovener (Stockholm)*, **53**, 251 (1973).
131. G. J. E. Balharry and D. J. D. Nicholas, *Anal. Biochem.*, **40**, 1 (1971).
132. A. Lundin and A. Thore, *Anal. Biochem.*, **66**, 47 (1975).
133. A. Lundin, A. Rickardsson, and A. Thore, *Anal. Biochem.*, **75**, 611 (1976).
134. P. E. Stanley, in *Organic Scintillators and Liquid Scintillation Counting*, edited by D. L. Horrocks and C. T. Peng, Academic Press, New York, 1971, pp. 607–620.
135. C. M. Jabs, W. J. Ferrell, and H. J. Robb, *Clin. Chem.*, **23**, 2254 (1977).
136. M. S. P. Manandhar and K. van Dyke, *Anal. Biochem.*, **58**, 368 (1974).
137. B. L. Strehler and W. D. McElroy, in *Method in Enzymology*, Vol. 3, edited by S. P. Colowick and N. O. Kaplan, Academic Press, New York, 1957, pp. 871–873.
138. D. M. Hercules and T. L. Sheehan, *Anal. Chem.*, **50**, 22 (1978).
139. P. E. Stanley, in ref. 71, p. 494.
140. A. Lundin and I. Styrelius, *Clin. Chim. Acta*, **87**, 199 (1978).
141. H. Zellweger and A. Antonik, *Pediatrics*, **55**, 30 (1975).
142. P. B. Addis and A. Antonik, in *2nd Biannual ATP-Methodology Symposium: Proceedings*, edited by G. Borun, SAI Technology Co., San Diego, 1977, p. 205.

143. S. A. Witteveen, S. E. Sobel, and M. De Luca, *Proc. Natl. Acad., Sci. USA*, **71**, 1384 (1974).
144. R. Fertel and B. Weiss, *Anal. Biochem.*, **59**, 386 (1974).
145. S. E. Brolin, E. Borglund, L. Tegner, and G. Wettermark, *Anal. Biochem.*, **42**, 124 (1971).
146. D. J. D. Nicholas and G. R. Clarke, *Anal. Biochem.*, **42**, 560 (1971).
147. C. Haggerty, E. Jablonski, L. Stav, and M. DeLuca, *Anal. Biochem.*, **88**, 162 (1978).
148. T. J. N. Carter, L. J. Kricka, and T. P. Whitehead, *Protides Biol. Fluid Proc. Colloq.*, **26**, 643 (1979).
149. E. W. Chappelle and G. L. Picciolo, *Methods Enzymol.*, **188**, 381 (1971).
150. P. E. Stanley, in *Liquid Scintillation Counting*, edited by M. A. Crook and P. Johnson, Heyden, London, 1974, pp. 253–271.
151. S. E. Brolin, G. Wettermark, and H. Hammar, *Strahlentherapie*, **153**, 124 (1977).
152. S. E. Brolin, *Bioelectrochem. Bioenerg.*, **4**, 257 (1977).
153. W. Cantarow and B. D. Stollar, *Anal. Biochem.*, **71**, 333 (1976).
154. E. Schram, R. Cortenbosch, E. Gerlo, and H. Roosens, in ref. 133, p. 125.
155. E. Jablonski and M. DeLuca, *Proc. Natl. Acad. Sci. USA*, **73**, 3848 (1976).
156. H. Hammar, *J. Invest. Dermatol.*, **65**, 315 (1975).

CHAPTER

3

# PRINCIPLE OF FLUORESCENCE MEASUREMENTS

## N. ICHINOSE AND G. SCHWEDT

Since the beginning of the 1970s, the spectrophotofluorometric method has been developed quickly, in parallel with the development of high-performance liquid chromatography. In the 1980s, different fluorometric apparatuses were manufactured and traded by a few factories—Shimazu, Hitachi, Nihonbunko, and so on—in Japan, and by about 20 factories in the Federal Republic of Germany. For convenience in handling their products, these manufacturers have described the physical measurement principles and the working points of the instruments in their instruction manuals. In addition, there are several monographes on fluorescent and phosphorescent spectroscopy by Guiltbault (1), Hercules (2), Parker (3), Schulman (4), Wehry (5) and Udenfriend (6); and a booklet by Perkin-Elmer (7) of the United States, as well as some papers (8–10) regarding the general measurements and synopsis of fluorescence. In this chapter, we discuss, from a side view of analytical chemistry, a newly important principle regarding the fluorometric instruments described in the original papers of quite recent years which was not covered in the above-mentioned monographs and other literature.

## 1. APPARATUS AND ARRANGEMENTS

A spectrophotofluorometer consists of four substantial components as follows;

1. Light source for excitation.
2. Monochromator or filter for the selection of excitation and emission wavelengths.
3. Measuring cell.
4. Photomultiplier as detector.

Tungsten and xenon lamps with a continuous beam and mercury and xenon lamps with a line beam are used in fluorometry, with a monochromator and a filter fluorometry, respectively, as the light source.

A monochromator with a grating makes it possible to take the spectra of fluorescence excitation and emission. A modern spectrophotometer frequently provides a supplementary optical and electric system and a reference photomultipler, at which the intensity of the light source holds constant.

Figure 3.1 shows the schematic construction of a spectrofluorometer. The light from the light source is focused through a first slit in front of the excitation monochromator or a prism filter in order to select the desired wavelength. The width of both slits, at the back of the excitation monochromator and in front of the emission monochromator, has a decisive influence on the resolving power of the spectrophotofluorometer: namely, the narrower the slit width, the larger is the selectivity or the resolving power; in brief, the closely coexistent signals on a chromatogram can be better distinguished. On the other hand, a smaller slit width reduces the intensity of light transmittance; thereby, the sensitivity also decreases. Further details on the significance of the slit are given in the report of Guilbault (1). The choice of an excitation or emission monochromator, respectively, plays an important role in the selection of a desired wavelength, as well as the determination of the bandwidth of a measured signal in conjunction with the slit width.

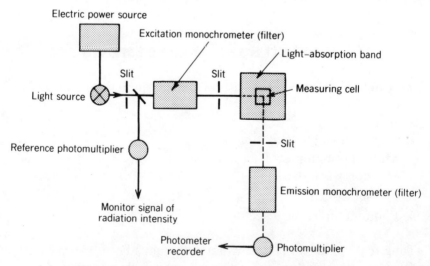

**Figure 3.1.** Schematic construction of a spectrophotometer.

As can be seen in Fig. 3.1, the excitation and emission radiations are at an angle of 90° to each other, in contrast to the situation with an absorption photometer. The photomultiplier, which indicates the intensity of absorption radiation by means of a photometer or recorder, serves as a radiation receiver.

### 1.1. Source of Light

Combination lamps [Hg-Xe, Hg-Cd, and Zn-Cd-Hg (11)], in addition to W, Xe, and Hg lamps, have been used as the light source for excitation (see Figs. 3.2 and 3.3).

Recently, some laser beams have been adopted as a new light source for excitation.

Requirements for the light source are as follows;

1. Strong intensity of the beam.
2. Equal distribution of the light intensity.
3. Temporal invariance of the spectral character and intensity.
4. Broad spectrum of the beam over UV and VIS emission regions.

Next, we describe briefly, some specifications of the light source in view of its applications to analytical chemistry.

1. *Tungsten lamp.* This is suitable for only the VIS area because of its spectral output and its usual encasement in a glass envelope which absorbs UV light.

2. *Xenon arc lamp.* This has a continuous spectrum from 250 to 650 nm (see Fig. 3.2) and an operating power of 100–150 W, and

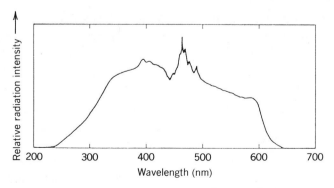

**Figure 3.2.** Emission spectrum of xennon lamp.

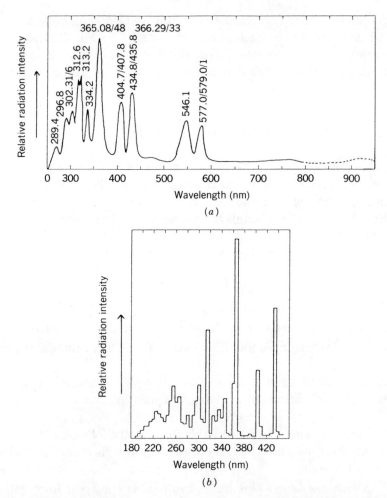

**Figure 3.3.** (a) Emission spectra of Hg and (b) the combined Zn–Cd–Hg lamps.

therefore is suitable for the "scanning" of spectra. This lamp has slightly poor stability; however, it can be regulated by using appropriate electrical equipment. A schematic construction of a double-beam fluorophotometer of the monitor type using the xenon arc lamp as light source is shown in Fig. 3.4.

3. *Mercury vapor lamp.* This has many intense lines over the range of 250–580 nm, as shown in Fig. 3.3; in practice, the maximum line of approximately 365 nm is used as an excitation light source. The spectral intensity distribution of the Hg lamp is dependent on the mercury vapor pressure when filled, and the lamp is classified accordingly as having high, mean, or low pressure. The low-pressure lamp

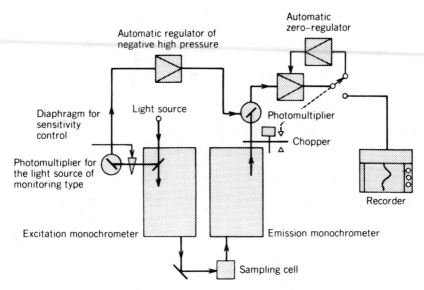

**Figure 3.4.** Schematic construction of a double-beam fluorophotometer in the light source of monitoring type.

radiates above all relatively intensive lines in its UV region; however, the effective range of wavelength is narrow. On the other hand, the mean- and high-pressure lamps mark a symmetrical intensity distribution of lines in both regions, UV and VIS. In particular, the mean-pressure lamp is suitable for excitation up to 365 nm; also, it has the longest duration of all Hg lamps.

4. *Mercury-xenon lamp.* This has more intense amplification than the Hg lamp.

5. *Mercury-cadmium lamp.* This has a wider emission band, ranging from 320 to 650 nm, than the Hg lamp.

6. *Laser.* This provides intense, very monochromatic radiation but can be applied only for an excitation range of narrowly limited wavelengths (e.g., double-frequency argon ionization laser (12): 257.9 nm; nitrogen UV laser: 337 nm). In general, lasers can be classified into four types: gas, liquid, and solid lasers, and semiconductors. Excimer lasers of $N_2$, Cu vapor, KrF, XeF, XeCl, and so on are used as practical gas lasers. The solid laser can be of Nd:YAG, glass, ruby, and so on; all these lasers have a narrow pulse width, with pulses from a few nanoseconds to scores of nanoseconds, and a large generating power, from 10 kW to scores of megawatts. The liquid laser, dye laser, is used in a photoexcitation system as well as the solid

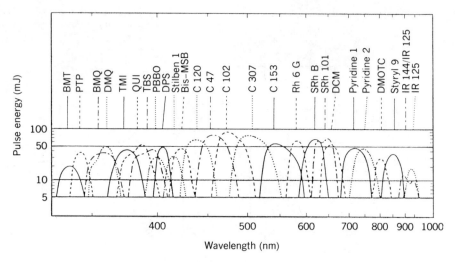

**Figure 3.5.** Dyes for excimer laser pumped dye lasers. (Type: EMG201E, tendered by Lambda Physik GmbH Co.).

laser; the dyes consist of inorganic or organic material. By using wide fluorescence spectra of 10–100 nm with the different dyes, it is possible to change continuously and synchronously the generating wavelengths, which radiate by photoexcitation, to the desired wavelength. In brief, the liquid laser is considered to be a wavelength convertor, which is able to change various wavelengths to a desired monochromatic wavelength of coherent light over the region of wavelengths from vacuum ultraviolet to near-infrared. As an example, the excimer-laser-pumped dye laser manufactured by Lambda Physik GmbH in the Federal Republic of Germany is shown in Fig. 3.5. The solid laser is characterized by high output power and amplification, even in a relatively small apparatus, and also by a very narrow photopulse up to a time width of $10^{-12}$ s. A semiconductor laser, "diode laser," is a form of the solid laser; it is now apparent that its value is remarkably enhanced by its high efficiency in regard to pulsing and the emitted output.

## 1.2. Monochromator and Filters

The desired wavelength for excitation or emission is selected by means of a filter fluorophotometer or fluorescence spectral photometer equipped with a monochromator.

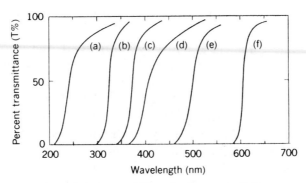

**Figure 3.6.** Permeability of various cutoff filters.

In a simplified filter fluorophotometer, various cutoff filters, which cutoff the light above or below a definite wavelength region are employed to select a desired spectral distribution of light (see Fig. 3.6). Alternative to the cutoff filter, a so-called band filter is allowed to pass only a relatively narrow spectral band. Guilbault (1) described in detail some common filters in the trade. A first filter, on the side of excitation, and a secondary filter, on the emission side, must be distinguished from each other on the basis of their filtering abilities. Generally, a cutoff filter is used as the first filter. For the secondary filter, higher resolving power than that of the first filter is required for the selection of the desired emission wavelength.

In analysis of a mixture, it is frequently necessary to use a narrow band filter in order to obtain adequate selectivity. In addition to the numerous cutoff filters, suitable bandpass filters to be placed over the spectral area of analytical interest in both UV and VIS regions have recently been placed on the market. These bandpass filters show high permeability of about 40% at bandwidths between 10 and 15 nm. Since the filter fluorophotometer has especially high sensitivity, it is used in combination with liquid or gas chromatography apparatus which gives superior selectivity. Apparatus equipped with monochromometers to analyze excitation and emission light are necessary for the analysis of mixtures or the determination of a specific refined structure in a substance, that is, the excitation and emission spectra are registered using the apparatus with each grating monochromator operated independently.

Cachelon (13) devoted himself to apparatus problems in monochromators. Mitchell et al. (14) reported on the fiber-optic filter fluorophotometer.

## 1.3. Measuring Cells

Pyrex glass (measurement at above 320 nm), quartz (measurement at UV), and fused silicon dioxide are used as materials for measuring cells in fluorophotometry. Since the measuring cell used in fluorophotometry absorbs only light that has a relatively narrow range of wavelengths, unlike the case in absorption photometry, the cell does not require stringent manufacturing conditions to maintain the accuracy of the fluorophotometry. In other words, the accuracy of fluorophotometry is not affected as much by the cell thickness and an error in the cell parallel plane at the transmitting face of light as would be the case in absorption photometry. However, it is an essential manufacturing condition of the cell that no material have any self-fluorescence. In fluorophotometry, a round cell is often used as the measuring cell, and in fact it is advisable to use a round cell, rather than a cuvette, in consideration of the accuracy of thickness and parallel plane at the transmitting cell walls. In fluorophotometric determination, unlike absorption photometry, fluorescence in a solution of the cell radiates in every direction, as shown in Fig. 3.7. However, in the practical determination of fluorescence in a cuvette, it is most appropriate to detect fluorescence that emits at right angles from the excitation light in order to limit the scattering light to a minimum (see Fig. 3.8). It is possible to measure fluorescence at 30° or 45° angles from the excitation light, but this is not usually done. A minimum sample volume for measurement in a microcuvette, by means of fluorophotometric analysis in connection with high-per-

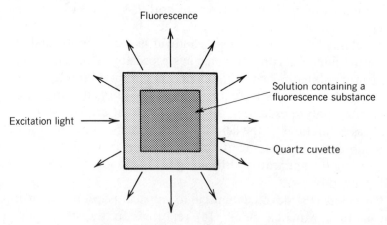

Figure 3.7. Divergence of fluorescence.

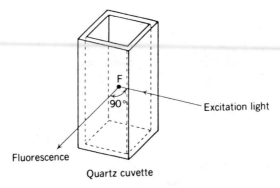

**Figure 3.8.** Fluorescence measurement using a cuvette. F is the fluorescence molecule.

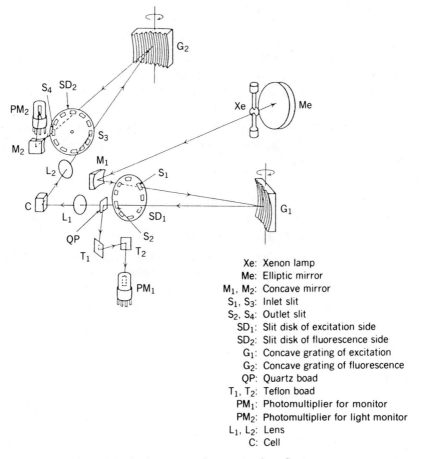

Xe: Xenon lamp
Me: Elliptic mirror
$M_1$, $M_2$: Concave mirror
$S_1$, $S_3$: Inlet slit
$S_2$, $S_4$: Outlet slit
$SD_1$: Slit disk of excitation side
$SD_2$: Slit disk of fluorescence side
$G_1$: Concave grating of excitation
$G_2$: Concave grating of fluorescence
QP: Quartz boad
$T_1$, $T_2$: Teflon boad
$PM_1$: Photomultiplier for monitor
$PM_2$: Photomultiplier for light monitor
$L_1$, $L_2$: Lens
C: Cell

**Figure 3.9.** Optic system of a spectrophotofluorometer.

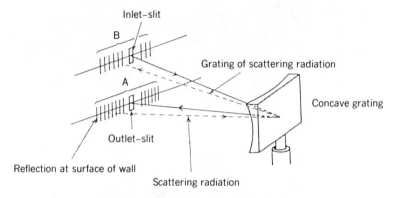

**Figure 3.10.** Off-plane spectrophotosystem.

formance liquid chromatography (HPLC), is about 5–20 μL. The excitation light in a fluorophotometer equipped with HPLC is focused through a lens or concave mirror in order to step down the central point of the excitation light, as shown in Figs. 3.9 and 3.10.

Rutili et al. (15) reported in detail on techniques regarding a measuring cell of $10^{-3}$-mm$^3$ sample volume and their applications to clinical chemical analysis. Smith et al. (16) described the construction and computation of the fiber-optic cell. Stieg and Nieman (17) reported on experimental and theoretical studies regarding the construction of the flowing cell in order to apply chemiluminescence to analytical chemistry.

### 1.4. Photomultiplier

A photomultiplier ("photomul"), secondary ion multiamplifier, is employed as fluorescence detector to obtain high sensitivity. The essential differences among individual photomultipliers lie in the materials used for the cathode and the window. An alkaline metal or its oxide in thin-phase form is a suitable cathode material for the photomultiplier because of its low ionization potential and the high electronic output power based on it. The photomultiplier cathode itself consists of silver/silver bromide. The composition of the material influences the maximum sensitivity and the spectral sensitivity distribution of the photomultiplier. Thus, the photomultiplier, which depends on its spectral sensitivity distribution, is classified on the basis of the photoelectric surface. The photoelectric surface possesses a maximum sensitivity in the wavelength range of 300–700 nm. In general, a photoelectric surface with Sb/Cs as a cathode material is

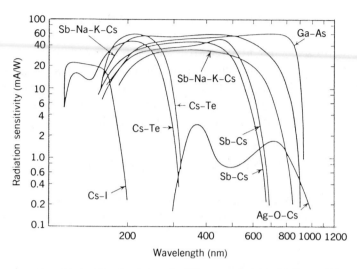

**Figure 3.11.** Sensitivity-distribution of different photomultipliers (Hamamatsu Photonics Co.).

employed. Figure 3.11 shows the spectral sensitivity distributions of different photomultipliers. The spectral region under 300 nm is also approachable when a quartz window is set in the photomultiplier instead of the conventional special glass window.

The sensitivity of fluorophotometric apparatus, which the efficiency of the photomultiplier determines, depends on so-called dark electricity. When a photon stops perfectly, a signal that occurs in a detector is due to the dark electricity, and its magnitude shows a temperature dependence. The emission signal/dark electricity proportion can be obtained from the ratio of the signal to its deflection. Improvement in the sensitivity or in the detection limit of analysis is achieved also by pulsed excitation-radiation, in addition to the technically difficult cooling of the photomultiplier. However, a critical report in regard to possible overestimation of this improvement has been written by Omenétto et al. (18). Ingle and Crouch (19) described the correlation obtained between modes and manufacturers of photomultipliers by means of the above-mentioned signal/deflection ratio.

It has been established from the spectral distribution of photomultipliers, obtained from experimental data shown in Fig. 3.11, that the sensitivities of these instruments show nonlinear relationships with wavelength. This fact means that photons with different energies produce different signal intensities at different wavelengths. These dependences are termed *receiver character*.

## 2. FLUORESCENCE MEASUREMENTS

Before a fluorescence analysis, it is necessary to determine the fluorescence excitation and fluorescence emission spectra of a desired fluorescent constituent in the sample, and then to select the suitable maximum wavelengths, that is, $\lambda_{ex}$ and $\lambda_{em}$. At the same time, it is necessary to run a blank test for fluorescence in the solvent used.

### 2.1. Fluorescence Emission Spectrum

Figure 3.12 shows an example of the fluorescence emission spectrum of an aqueous solution of salicylic acid (10 ppm) when the fluorescence emission wavelength was scanned from 250 nm to longer wavelengths at $\lambda_{ex}$ of 280 nm.

When the fluorescence emission spectrum in a dilute aqueous solution is scanned, the overlapping of variously interfering excitation spectra, as well as the desired fluorescence spectrum, is observed. Consequently, it is necessary to remove these interfering spectra from the desired spectrum in order to obtain a true fluorescence emission spectrum. In the example in Fig. 3.12, the fluorescence emission spectrum of sodium salicylate in an aqueous solution lies on top of scattering excitation light (Rayleigh scattering), which arises from scattering molecules and dust in the solvent, the Raman spectrum of impurities in the solvent and in the cell, and secondary scattering of

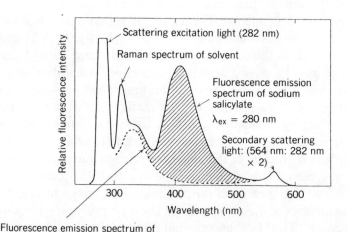

**Figure 3.12.** Fluorescence emission spectrum of sodium salicylate.

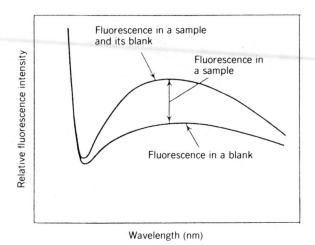

**Figure 3.13.** Underfluorescence of a blank.

light. Peak areas other than those of the shaded parts in Fig. 3.12 indicate an underfluorescence of the blank (see Fig. 3.13). The first scattered excitation light among the chromatographs in Fig. 3.12 has the same wavelength as the excitation light, 282 nm, because of the signal that occurred as a result of reflection of the excitation light from dust and bubbles in the solvent. It is possible to make this light very low by minimizing the slit width in the side of the excitation light, although it cannot be completely eliminated because of the remaining Rayleigh scattering.

Next, the Raman spectrum of a solvent, appearing near the above-mentioned scattered light, arises from Raman activity by the solvents water, benzene, and so on; it is very difficult to distinguish from the fluorescence emission spectrum. However, it is possible to differentiate the spectra by utilizing the shift of the Raman spectrum accompanying a wavelength change of the excitation light, since the scattered light and the Raman spectrum always appear at a constant energy distance from each other, according to Stokes law. On the other hand, when concentration of a desired constituent increases, Raman scattered radiation decreases. On the basis of this experimental fact, the two spectra can be distinguished (see Fig. 3.14).

When a grating is used, a small peak of secondary scattered light appears at just double the distance of the scattered light. This secondary scattered light can be excluded simply by inserting an excitation cutoff filter in front of the fluorescence monochromator. Using a double monochromator—grating type and excitation monochromator—causes such secondary scattering light to disappear.

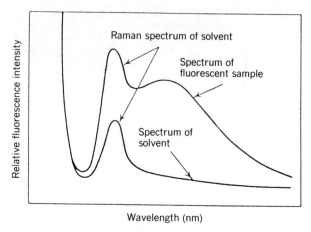

Figure 3.14. Raman-scattering radiation of a solvent.

## 2.2. Fluorescence Excitation Spectrum

Figure 3.15 shows a fluorescence excitation spectrum of an aqueous salicylate solution when it is scanned from 200 nm to longer wavelengths at $\lambda_{em}$ of 405 nm, obtained from the experimental results in Fig. 3.12. It can be seen from this experimental result in Fig. 3.15 that a more sensitive fluorescence emission spectrum of salicylate aqueous

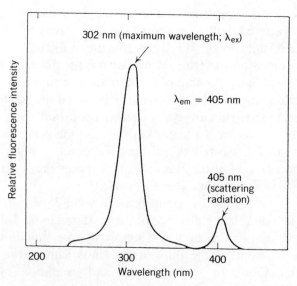

Figure 3.15. Fluorescence excitation spectrum of sodium salicylate.

solution is obtained by altering the first fluorescence excitation of 280 nm, used in Fig. 3.12, to 302 nm. The maximum wavelength of the fluorescence excitation spectrum agrees very closely with that of its absorption spectrum; therefore, when the fluorescence emission spectrum of an unknown, desired component is initially scanned, use of a maximum wavelength of its absorption spectrum as $\lambda_{ex}$ is recommended.

## 2.3. Calibration Curve of Fluorescence

The relationship between fluorescence intensity and concentration is shown in Eqs. (3.1) and (3.2).

The intensity of fluorescence that radiates from a point in the cell conforms to Eq. (3.1):

$$dB(\lambda') = \frac{\rho}{4\pi n^2} E\lambda F\lambda(\lambda') K\lambda \, d\lambda' \tag{3.1}$$

where $dB(\lambda')$ = fluorescence intensity observed at wavelength $\lambda'$;
$n$ = refractive index;
$\rho$ = reflection coefficient;
$E\lambda$ = intensity of excitation light at wavelength $\lambda$;
$F\lambda(\lambda')$ = true fluorescence intensity at wavelength $\lambda'$ among spectra radiated by the excitation light at wavelength $\lambda$;
$K\lambda$ = absorption at wavelength $\lambda$;
$d\lambda'$ = bandwidth at wavelength $\lambda'$.

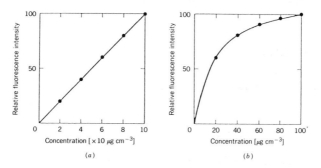

**Figure 3.16.** Calibration curves of diaminostilbene aqueous solution.

Then the absorbance is in proportion to concentration, $C$; therefore, Eq. (3.2) is obtained by integrating Eq. (3.1):

$$B(\lambda') = KC \qquad (3.2)$$

Since $dB(\lambda')$ is proportional to $C$ by Eq. (3.2), the calibration curve of $dB(\lambda')$ versus $C$ gives a straight line. However, when the concentration of a desired component increases too much, the calibration curve gives a nonlinear plot because fluorescence of relatively short wavelength, which radiates from the center of the sample, is absorbed again in the sample solution; that is, "self-absorption" occurs. Calibration curves of diaminostilbene aqueous solution are shown in Fig. 3.16 as examples of the above-mentioned cases.

## REFERENCES

1. G. G. Guilbault, *Practical Fluorescence—Theory, Methods and Techniques*, M. Dekker, New York, 1973.
2. D. M. Hercules, *Fluorescence and Phosphorescence, Analysis Principles and Applications*, Interscience, New York, 1966.
3. C. A. Parker, *Photoluminescence in Solutions with Applications to Photochemistry and Analytical Chemistry*, Elsevier, New York, 1968.
4. S. G. Schulman, *Fluorescence and Phosphorescence Spectroscopy*, Pergamon, Oxford, 1977.
5. E. L. Wehry, *Modern Fluorescence Spectroscopy*, Vols. 1 and 2, Plenum Press, New York, 1976.
6. S. Udenfriend, *Fluorescence Assay in Biology and Medicine*, Vol. 2, Academic Press, New York, 1969.
7. Perkin-Elmer (Bodenseewerk), *Einführung in die Fluoreszenz-Spektroskopie*, Perkin-Elmer, Überlingen, 1978.
8. B. L. Van Duuren and T. L. Chan, *Advan. Anal. Chem. Instrumen.*, **9**, 387 (1971).
9. P. F. Lott and R. J. Hurtubise, *J. Chem. Educ.*, **51**, A315–320, A358–364 (1974).
10. M. A. West, *Am. Lab.*, **7**, 57 (1975).
11. J. T. H. Goosen and J. G. Kloosterboer, *Anal. Chem.*, **50**, 707 (1978).
12. J. S. Shirk and A. M. Bass, *Anal. Chem.*, **41**, 103 (1969).
13. M. Cachelon, *Method. Phys. Anal.*, **5**, 298 (1969).
14. D. G. Mitchell, J. S. Garden, and K. M. Aldons, *Anal. Chem.*, **48**, 2275 (1976).

15. G. Rutili, K. E. Arfors, and H. R. Ulfendahl, *Anal. Biochem.*, **72**, 539 (1976).
16. R. M. Smith, K. W. Jackson, and K. M. Aldous, *Anal. Chem.*, **49**, 2051 (1977).
17. S. Stieg and T. A. Nieman, *Anal. Chem.*, **50**, 401 (1978).
18. N. Omenétto et al., *Anal. Chem.*, **49**, 1076 (1977).
19. J. D. Ingle and S. R. Crouch, *Anal. Chem.*, **43**, 1331 (1971).

CHAPTER

4

# BIOCHEMICAL AND BIOMEDICAL APPLICATIONS

## 1. BIOCHEMISTRY

F. -M. SCHNEPEL

### 1.1. Amines

#### *1.1.1. Native Fluorescence*

Some amines, in which the nitrogen atom is part of an aromatic system (e.g., pyrimidines, purines, indoles, etc.; see Fig. 4.1), can be determined by means of their native fluorescence. Fluorescence spectrometry has been applied to purines and pyrimidines by Udenfriend and Zaltzman (1). The results show that some compounds emit appreciable fluorescence: guanine fluorescence is sufficiently intense and specific to be used for assay in nucleic acid hydrolysates and other mixtures. Methylaminoguanine and dimethylaminoguanine fluoresce even more intensely. Fluorescence measurements can be used as a very sensitive procedure for determining the purity of purines and pyrimidines; several commercially available compounds, which seemed to be "pure" by other criteria, contained sufficient impurities to distort the excitation and fluorescence spectra.

The native fluorescence of indoles, too, is sufficient for quantitative determination purposes. Indoles that are not substituted at the ring system fluorescece at 348 nm [excitation wavelength 287 nm; corrected values reported by Teale and Weber (2)]. The determination of substituted 2,3-bis(4-methoxyphenyl) indoles in serum, urine, and feces is described by Kaiser et al. (3). The method is based on ethyl acetate extraction of alkaline specimens and is sensitive to 0.1–0.7 $\mu$g/100 mg sample.

Some spectral data concerning the native fluorescence of amines are summarized in Table 4.1.

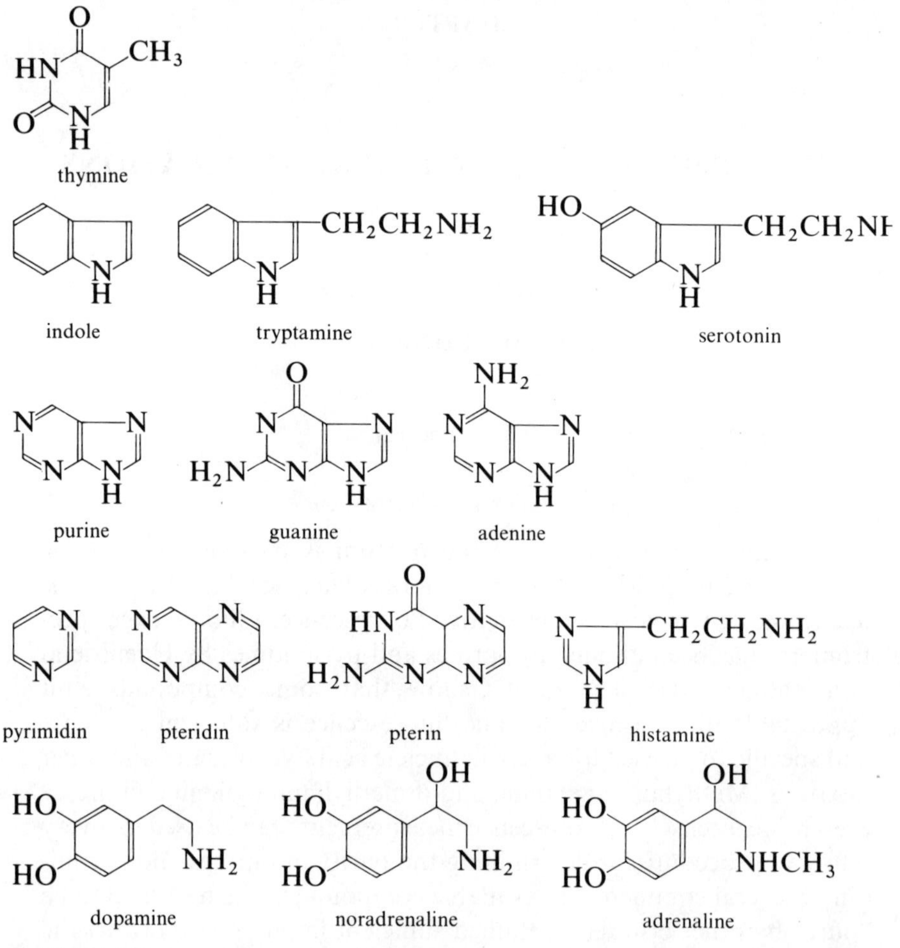


| | |
|---|---|
| $H_2N(CH_2)_4NH_2$ | putrescine |
| $H_2N(CH_2)_5NH_2$ | cadaverine |
| $H_2N(CH_2)_4NH(CH_2)_3NH_2$ | spermidine |
| $H_2N(CH_2)_3NH(CH_2)_4NH(CH_2)_3NH_2$ | spermine |

**Figure 4.1.** Structure formula of some amines.

### *1.1.2. Derivatizations*

For the determination of amino groups, several organic reagents have been described (see Tables 4.2 and 4.3). Some reactions and their applications will be discussed in more detail.

**1.1.2.1. Compounds with $NH_2$ Groups.** A convenient method for the determination of primary amines is the reaction with fluorescamine (Fig. 4.2*a*). The conditions and the principal parameters affecting this

**Table 4.1. Fluorescence Data Concerning the Native Fluorescence of Some Amines**

| Compound | Wavelengths (nm) | |
|---|---|---|
| | Excitation | Emission |
| Aromatic amines | | |
|   Catecholamines | 285 | 325 |
|   $o$-Methylcatecholamines | 285 | 335 |
|   Tryptamine | 285 | 360 |
|   Indolamines | 287 | 348 |
| Indoles | 320–330 | 415–420 |
| Purines, pyrimidines and coenzymes | | |
|   Purine | 285 | 370 |
|   Adenine | 265 | 380 |
|   Adenosine, ADP, ATP | 272 | 390 |
|   Guanine | 275 | 360 |
|   Isoquanine | 300 | 360 |
|   Thymine | 290 | 380 |
|   NADH | 340 | 460 |

reaction were reexamined by Castell et al. (15). The influence of steric hindrance on the derivatization of aromatic amines was studied by Tomkins et al. (16). Sterically unhindered monocyclic primary aromatic amines and 2-naphthylamine form stable, fluorescent derivatives with fluorescamine in both aqueous and nonaqueous solutions. Aromatic amines with a substituent such as a methyl group or an aromatic ring ortho to the amine functional group derivatized either poorly or not at all.

A review of the analytical application of fluorescamine to the identification of biologically important primary amines was published by Kusnir (14).

For the fluorescent detection of picomole quantities of primary amines, the reagent $o$-phthalaldehyde is currently used in combination with a thiol compound, usually $\beta$-mercaptoethanol (see Fig. 4.2b). Using ethanethiol instead of $\beta$-mercaptoethanol increases both the stability of the fluorescent product and its spectral responses to changes in solvent polarity, without appreciably altering the observed fluorescence intensity (31). For further details of the reaction of

Table 4.2. Derivatization of Amines, Amino Acids, and Related Compounds

| Compound | Reagent | Reference |
|---|---|---|
| Primary amines | Salicylaldehyde-diphenylborane chelate | 4 |
| Primary aliphatic amines | 1-Pyrenealdehyde | 5 |
|  | 2-Fluorenealdehyde | 5 |
| Primary and secondary amines | 4-(6-Methylbenzothiazol-2-yl)phenylisocyanate | 6 |
| Primary and secondary aliphatic amines | 9-Isothiocyanatoacridine | 7 |
| Secondary amines | 2-Methoxy-2,4-diphenyl-3(2$H$)-furanone | 8 |
| Tertiary amines | 9,10-Dimethoxyanthracene-2-sulfonate | 9, 10 |
| Primary amines, amino acids | Fluorescamine | 11–16 |
|  | 5-Isothiocyanato-1,8-naphthalenedicarboxy-4-methylphenylimide | 17 |
| Amines, amino acids | Dansyl chloride | 18–21 |
|  | Bansyl chloride | 22 |
|  | NBD-chloride | 23 |
|  | Ninhydrin | 24 |
|  | Pyridoxal | 25 |
|  | Fluorescein-isothiocyanate | 26 |
|  | 2,3-Diphenyl-1-indanone | 27 |
|  | 2,4-Butandione | 28 |
|  | $o$-Phthalaldehyde | 29–31 |

| | | |
|---|---|---|
| Peptides | o-Phthalaldehyde | 32, 33 |
| | Ninhydrin | 24 |
| | Dansyl chloride | 34 |
| | Bansyl chloride | 22 |
| | Fluorescamine | 35 |
| Proteins | Fluorescamine | 15, 35 |
| | Fluorescein-isothiocyanate | 26 |
| | 4-Dimethylamino-4'-isothiocyanostilbene | 36 |
| | 2-Toluidene-naphthalene-6-sulfonic acid | 37 |
| Polyamines | Dansyl chloride | 34 |
| | o-Phthalaldehyde | 38, 39 |
| Biogenic amines | o-Phthalaldehyde | 40–42 |
| Carbamates | NBD-chloride | 43 |
| Acrylamines | N-(1-naphthyl)ethylendiamine | 44 |
| Nitrosamines | NBD-chloride | 45, 46 |
| Amino sugars | Ninhydrin | 24, 47 |
| | Pyridoxal | 48 |

## Table 4.3. Steps in the Derivatization of Amines

| Compound | Step in Derivatization | Reference |
|---|---|---|
| Primary and secondary amines | 1. Reaction with 1,2-naphthoquinone-4-sulfonic acid<br>2. Reduction with $KBH_4$ | 49, 50<br><br>51 |
| Primary aromatic amines | 1. Diazotization of the amino group<br>2. Coupling with 2,6-diaminopyridine or 4-methoxy-*m*-phenylenediamine<br>3. Reaction with copper ions in alkaline solution | 52<br><br>53 |
| Biogenic amines | | |
| a. Adrenaline, noradrenaline, 3,4-dihydroxyphenyl-alanine, normetanephrine, metanephrine | 1. Formation of adrenochromes by oxidation<br>2. Formation of trihydroxyindole derivatives by reduction and isomerization in alkaline solution (see Fig. 4.5*a*) | 54–56 |
| b. Adrenaline, noradrenaline, dopamine | 1. Formation of adrenochromes by oxidation<br>2. Condensation with ethylenediamine (see Fig. 4.5*b*) | 57 |
| c. Tryptamine | 1. Reaction with formaldehyde<br>2. Formation of norharman by oxidation (see Fig. 4.6) | 58 |

primary amino groups with *o*-phthalaldehyde see Section 1.2 (amino acids).

1-Pyrenealdehyde and 2-fluorenealdehyde are fluorogenic reagents for primary aliphatic amines, forming Schiff's bases (see Fig. 4.2c) that are intensively fluorescent in acidic ethanol. Derivatization with 1-pyrenealdehyde has several advantages; for example, the derivatized amines can be detected at concentrations about one order of magnitude lower than those of 2-fluorenealdehyde derivatives, the reagent is cheaply available in a state of reasonable purity, and it is effective specifically for primary aliphatic amines. Combination with a depro-

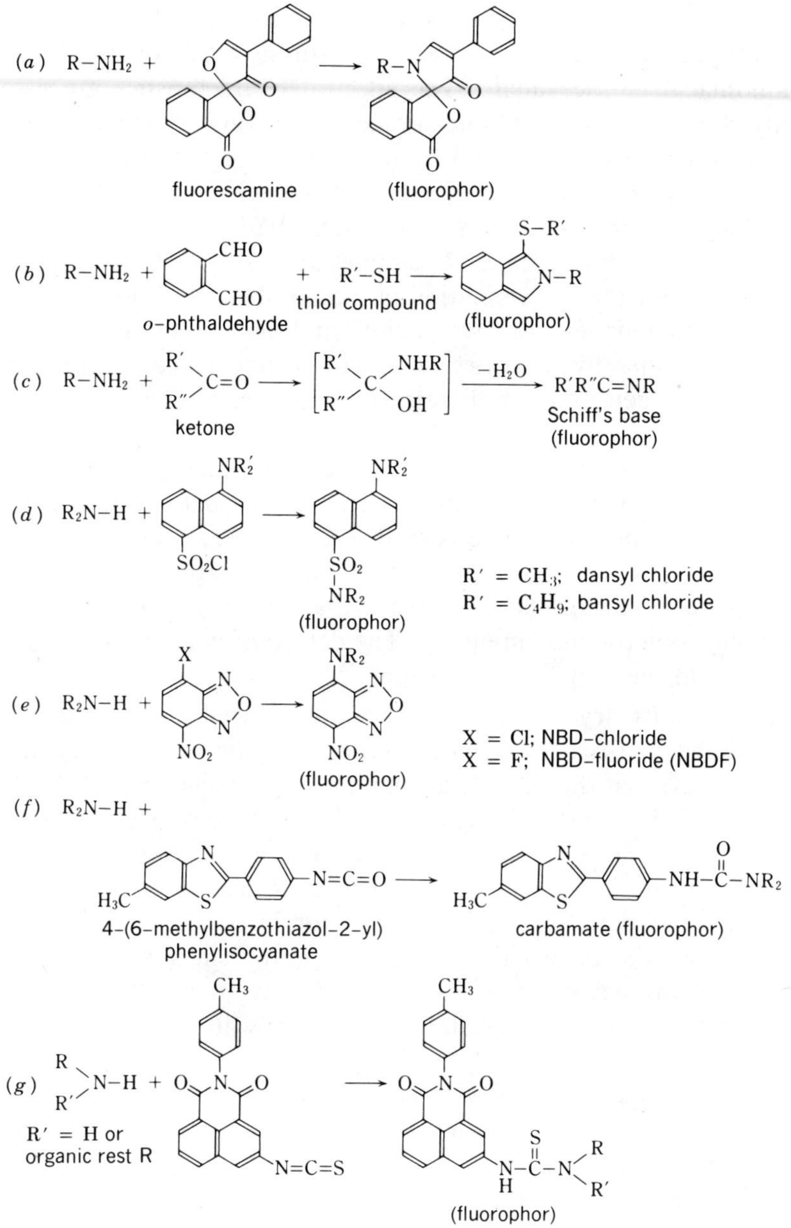

Figure 4.2. Some derivatization reactions of primary and secondary amines.

teinizing step permits the determination of primary amines in blood serum in nanogram quantities (5).

A sensitive fluorometric method for primary aromatic amines is reported by Dombrowski and Pratt (52). The procedure involves diazotization of the amine, followed by coupling with 2,6-dimethylaminopyridine and reaction of the resulting azo compound with ammoniacal cupric sulfate. The spectral characteristics are essentially common among various amines (excitation, 360 nm; emission, 420 nm). The amines can be determined in the region of 2–6 ng cm$^{-3}$.

A similar method is described by Taniguchi et al. (53). For the coupling reaction, 4-methoxy-*m*-phenylenediamine is used instead of 2,6-dimethylaminopyridine. A simple and sensitive procedure is established for the determination of *p*-aminobenzoic acid.

Two methods for the separation and identification of several primary amines by derivatization, thin-layer chromatography, and fluorometric detection are described by Hohaus (4, 59):

1. By reaction with excess salicylaldehyde-diphenylboron chelate, primary amines form fluorescent salicylaldehyde-azomethine-diphenylboron chelate (see also Chapter 5, Section 3). These derivatives are dissolved in chloroform and applied to thin-layer plates of silicagel 60. The chromatograms are evaluated by *in situ* fluorometric measurements. The detection limits of this method are in the range of 4–6 ng mm$^{-3}$.

2. The primary alkylamines are derivatized into 3-alkyl-2,2-diphenyl-1-oxa-3-azonia-2-boratanaphthalenes and separated by reversed thin-layer chromatography on silanized silicagel-60 plate. As mobile phase, an 80 : 20 mixture of MeOH–H$_2$O is applied. The quantitative analysis of amine mixtures is achieved by *in situ* fluorometry of the developed chromatogram. With a limit of detection of 4 ng mm$^{-3}$ for hexylamine, this method allows the determination of 20 ng mm$^{-3}$ in a sample with a standard deviation of ± 1 ng mm$^{-3}$ hexylamine. The separation and detection of decylamine from river water are described.

**1.1.2.2. Compounds with NH$_2$ or NHR Groups.** Common methods are the reactions with 5-dimethylaminonaphthalene-1-sulfonyl chloride (dansyl chloride), 5-dibutylaminonaphthalene-1-sulfonyl chloride (bansyl chloride), or 7-chloro-4-nitrobenzo-2-oxa-1,3-diazole (NBD-chloride); the reactions and the reaction products are shown in Fig. 4.2*d*, *e*. An outline of the various methods that use dansyl chloride, with emphasis on the detection of amines and amino acids in biological tissues, is presented by Leonard and Osborne (19).

Wintersteiger et al. (6) report another derivatization method for the quantitative determination of primary and secondary amines. In the presence of an equimolar amount of triethylenediamine as catalyst, the compound 4-(6-methylbenzothiazol-2-yl)-phenylisocyanate reacts with amines to form highly fluorescent derivatives (see Fig. 4.2f). Primary aliphatic amines react at room temperature within 5 min and secondary aliphatic amines within 10 min, whereas heterocycle secondary amines require more time. After excess isocyanate has been destroyed with diethylamine, the method can be combined with thin-layer chromatography or HPLC. By means of thin-layer chromatography, 50 pg of amine can be detected as fluorescing carbamate. The same derivatization method can be applied to alcoholic compounds, which form highly fluorescing urethanes with the reagent.

Primary and secondary amines can also be derivatized to fluorescent products with the reagent 5-isothiocyanato-1,3-dioxo-2-$p$-tolyl-2,3-dihydro-1$H$-benz(de)isoquinoline [Khalaf and Rimpler (60)]. The amines are reacted at 60°C with the reagent to form fluorescent thiourea compounds (see Fig. 4.2g) with excitation maxima at 425–430 nm and emission maxima at 475–480 nm. Amino acids are derivatized at 90°C to thiocarbamyl amino acids (see Section 1.2).

Compounds with secondary amino groups can also be determined fluorometrically by reaction with fluorescamine (see Fig. 4.3a) at pH 12 (61). After the reaction with fluorescamine, the solution is heated at 70°C for 10 min with primary amine (L-Leu-L-Ala) and thereby exhibits a bluish green fluorescence. By measurement of the fluorescence, most secondary amines could be determined in 0.5-nmol quantities. Relative standard deviations were 4.4 and 5.2% for the analyses of 5 nmol $N$-methylaniline and sarcosine, respectively.

Similarly to fluorescamine, the analogous compound 2-methoxy-2,4-diphenyl-3(2$H$)furanone (MDPF) reacts with primary and secondary amines (8). Primary amines form fluorescent derivatives, whereas the products of secondary amines are nonfluorescent but can react with a primary amine (e.g., taurine) to form fluorescent compounds (see Fig. 4.3b). The secondary amines are reacted with MDPF at pH 10 and 20°C for 45 min and then incubated at pH 9 and 45°C for 10 min with taurine. By measurement of the bluish green fluorescence (wavelength of excitation, 390 nm; of emission, 480 nm), several secondary amines can be determined at micromole per cubic meter levels. Compared to the analogous method with fluorescamine (see above), however, the reactivity of MDPF with secondary amines is lower, and the method seems to be less sensitive.

Another fluorometric method for the determination of secondary amines is discussed by Himuro et al. (62). The method is based on

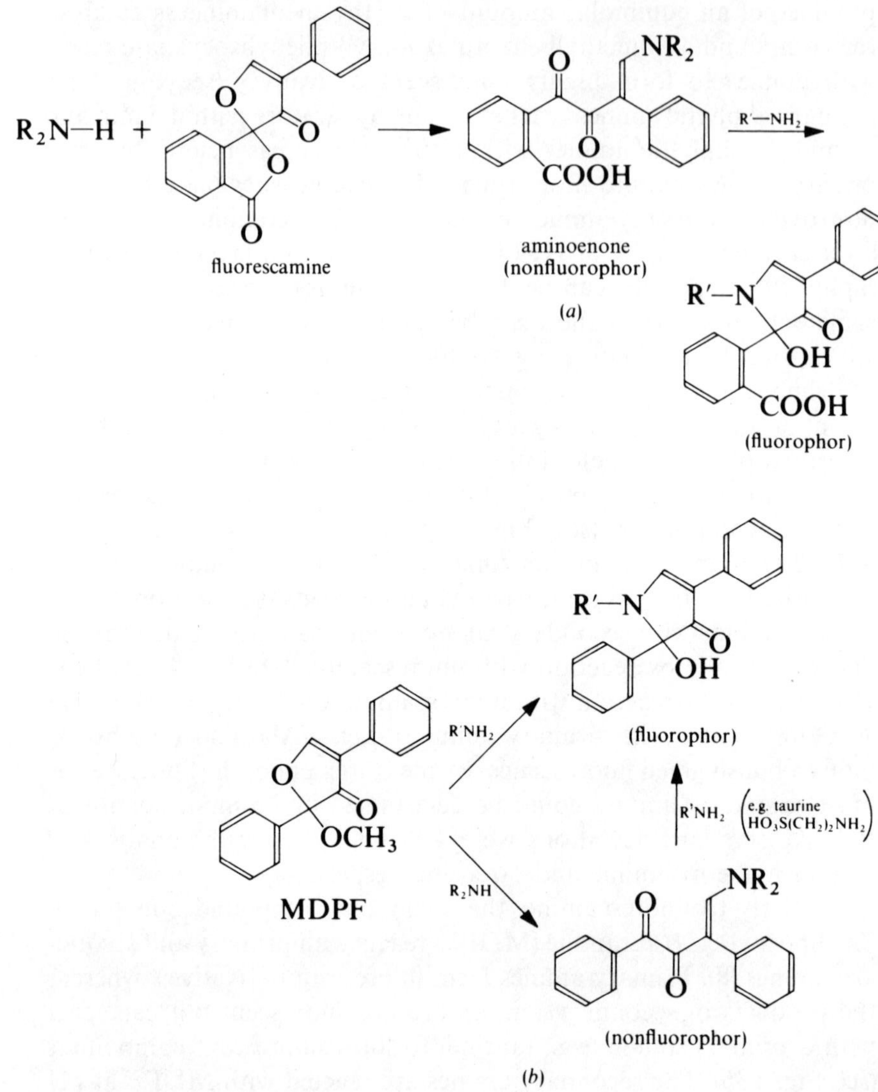

**Figure 4.3.** (a) Reaction of fluorescamine with secondary amines, and (b) reaction of MDPF with primary and secondary amines.

conversion of the secondary amines with NaClO to primary amines (see Fig. 4.4a), which are then reacted with the o-phthalaldehyde-mercaptoethanol reagent (see Fig. 4.2b). The amount of secondary amines in the presence of primary amines can be determined by comparing fluorescence intensities with and without hypochlorite treatment.

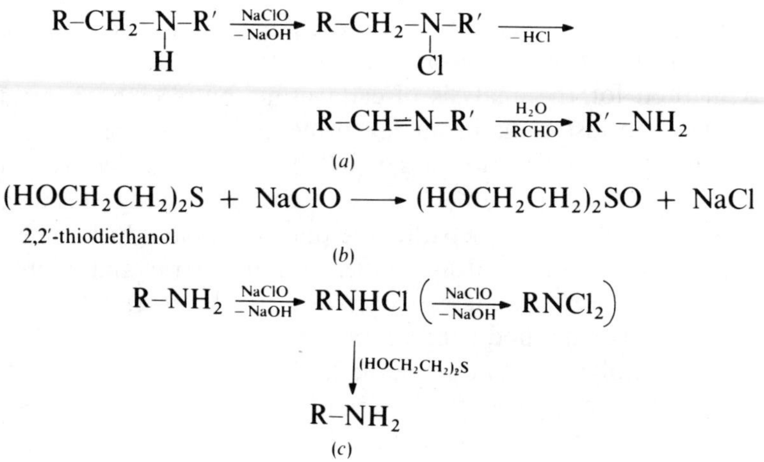

$$R-CH_2-\underset{H}{N}-R' \xrightarrow[-NaOH]{NaClO} R-CH_2-\underset{Cl}{N}-R' \xrightarrow{-HCl}$$

$$R-CH=N-R' \xrightarrow[-RCHO]{H_2O} R'-NH_2$$
(a)

$$(HOCH_2CH_2)_2S + NaClO \longrightarrow (HOCH_2CH_2)_2SO + NaCl$$
2,2'-thiodiethanol
(b)

$$R-NH_2 \xrightarrow[-NaOH]{NaClO} RNHCl \left( \xrightarrow[-NaOH]{NaClO} RNCl_2 \right)$$

$$\downarrow {\scriptstyle (HOCH_2CH_2)_2S}$$

$$R-NH_2$$
(c)

**Figure 4.4.** (a) Reduction of secondary amines by NaClO, and (b, c) reactions of 2.2'-thiodiethanol.

Interferences by excess hypochlorite (e.g., oxidation of the reagent or conversion of primary amines to chloramines) were suppressed by 2,2-thiodiethanol (see Fig. 4.4b, c). Two spectrofluorometric methods are reported, one for aromatic and the other for aliphatic secondary amines. These methods permitted detection at nanomolar levels; relative standard deviations were 4% for 5 nmol N-methylaniline and 13% for 5 nmol sarcosine.

**1.1.2.3. Compounds with $NR_3$ Groups.** Sodium coumarin-6-sulfonate was used as fluorescent ion pair reagent for the analysis of tertiary amines by Dent et al. (63). $CH_2Cl_2$ containing 5% n-pentanol was used to extract the coumarin-amine ion pair from aqueous solution. After phase separation, the coumarin species was completely ionized by the addition of tetrabutylammonium hydroxide to the organic phase. After irradiation for 30 min by long-wavelength UV light (365 nm), the fluorescence intensity of the sample was measured using excitation and emission wavelengths of 400 and 540 nm, respectively. This method was applied to the determination of chlorpheniramine maleate with an accuracy of 4–6% and a precision of 2–6% relative standard deviation. The calibration curve was linear in the 50–100 ng cm$^{-3}$ (0.13–2.6 × 10$^{-6}$ mol dm$^{-3}$) range. Detection limits were 50 ng cm$^{-3}$ (35.3 ng cm$^{-3}$ of chlorpheniramine free base).

**1.1.2.4. Polyamines.** An improved method for the routine analysis of polyamines in biological fluids with a conventional amino acid analyzer was described by Milano et al. (39). The method includes the use of a sulfonated polystyrene cation-exchange resin and continuous fluorescent detection with o-phthalaldehyde.

A method for the analysis of cadaverine, putrescine, spermidine, spermine, and histamine in human urine and animal cell extracts is described by Andrews and Baldar (64). This assay, which uses automated chromatography on Ultrapac 8 cation-exchange resin in an amino acid analyzer, can separate the polyamines and histamine with a single, mixed Na/K citrate buffer. Fluorometric and ninhydrin methods were used for analyses of hydrolyzed urine and cell extract, respectively. The method is rapid and sensitive and allows the determination of polyamines at picomole levels.

A fluorometric assay for total diamines in human placental diamine oxidase is described by Matsumoto et al. (65). The diamines were purified from the urine by cation-exchange chromatography and incubated with human placental diamine oxidase. $H_2O_2$ was formed in the diamine oxidase reaction and measured fluorometrically by converting homovanillic acid to a highly fluorescent compound in the presence of peroxidase. Because of its simplicity and high sensitivty, the method seems useful for routine clinical investigations. The data obtained from normal subjects and from patients suffering from various forms of cancer are also presented.

The determination of urinary di- and polyamines, including spermine, after separation by thin-layer chromatography is described by Beyer and Van den Ende (66). After acid hydrolysis of the samples, the amines are isolated by cation-exchange chromatography, derivatized to dansyl products in a water bath at 70°C for 30 min, and extracted with PhMe. Ascending TLC is carried out on activated silica gel plates with a mixture of $CHCl_3$–PhMe–$Et_3N$ as developing agent. The amines are quantitated by fluorometry.

**1.1.2.5. Biogenic Amines.** A review of fluorescence microscopy and microfluorometry of biogenic monoamines was presented by Lindvall et al. (67). The paper includes the chemistry of the fluorophor-forming histochemical reactions, and variations of HCHO and glyoxylic acid methods for visualization of biogenic amines.

An important derivatization of catecholamines is the formation of trihydroxyindole compounds (see Fig. 4.5a). In the first step of reaction, adrenaline and noradrenaline are oxidized by iodine or potassium hexacyanoferrate(III) to form so-called adrenochromes, which in the second step are isomerized in alkaline solution to trihydroxyindole derivatives. The fluorophors are stabilized by addition of acid to pH 5. A review of this method was published by Wisser and Knoll

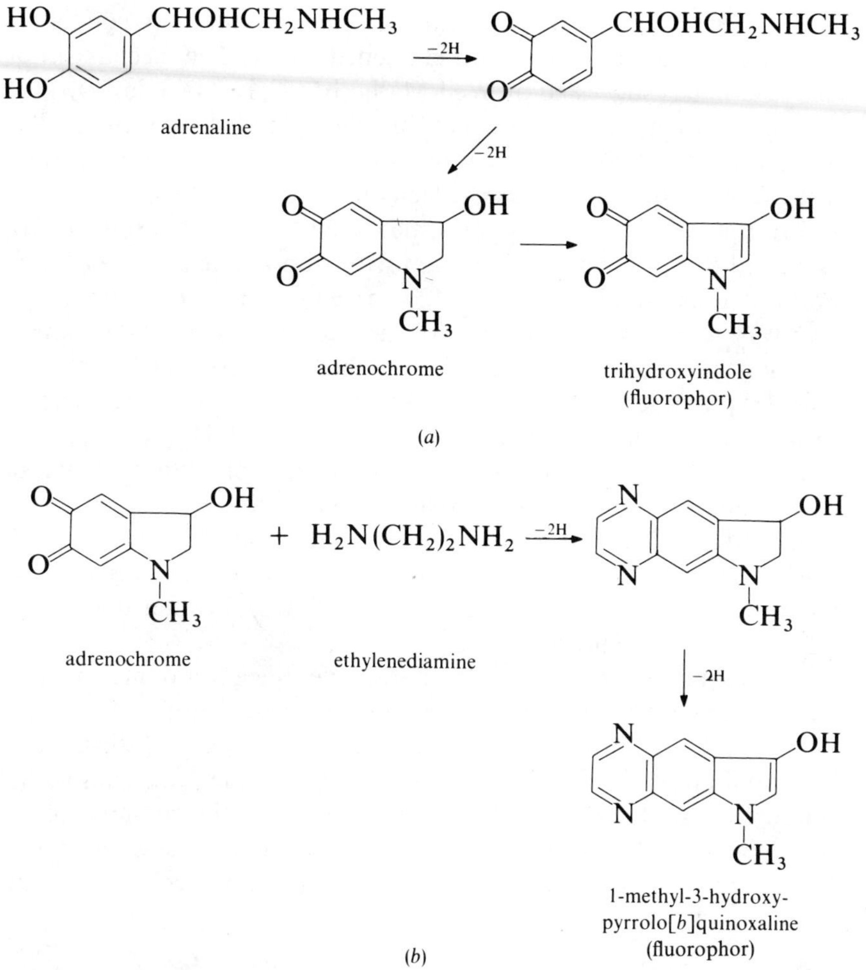

**Figure 4.5.** Derivatization of adrenaline; (a) formation of a trihydroxyindole, (b) condensation with ethylenediamine.

(56). The effects of different reducing agents (e.g., ascorbic acid, sodium sulfite, 2-mercaptoethanol) on stabilization of the fluorophors were used for more refined analyses of adrenaline and noradrenaline (68, 69).

The o-methylmetabolites of the catecholamines, metanephrine and normetanephrine, are determined in the same way (40, 70, 71), and dopamine, 3-dopamine, and 3-methoxytyramine can also be analyzed fluorometrically after derivatization to dihydroxy indole compounds, which are formed by the same mechanism (40).

Another important derivatization of catecholamines is condensation of the adrenochromes with ethylenediamine. The mechanism of this reaction as suggested by Weil-Malherbe (57) is shown in Fig. 4.5b. As ethylenediamine reacts with all catechols, several separation steps are necessary to use this reaction as a relatively specific procedure. Therefore, acidic and nonbasic catechols are removed by ion exchange before the derivatization is performed. The fluorescent products are extracted from the aqueous solution with isobutanol, and their fluorescence is measured (excitation, 420 nm; emission, 510–580 nm).

A very selective derivatization for the fluorometric determination of the biogenic amine tryptamine is described by Hess and Udenfriend (58). Tryptamine reacts with formaldehyde in the solvent benzene to form tetrahydronorharmane, which is oxidized by $H_2O_2$ to norharmane (see Fig. 4.6). This product is fluorescent with maxima of excitation at 365 nm and of emission at 440 nm.

A modified fluorometric method for the determination of serotonin (5-HT), dopamine, and norepinephrine in brain tissue is presented by Ciarlone (41, 42). It yields recoveries of 100% 5-HT after derivatization with o-phthalaldehyde.

Techniques for the microfluorometric identification of monoamine fluorophors are described by Reinhold and Hartwig (72). By extension of the excitation range to 250 nm, it is possible to distinguish fluorophors of DOPA from those of dopamine, fluorophors of noradrenaline from those of adrenaline, and fluorophors of 5-hydroxytryptophan from those of 5-HT. In a computer-assisted on-line scanning procedure using a high-aperture inverted illuminating

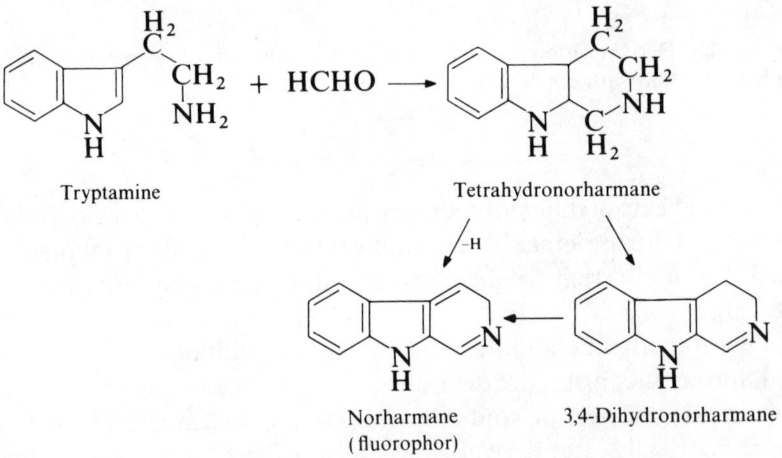

**Figure 4.6.** Derivatization of triptamine.

microscope equipped with UV-transmitting optics, the specific formaldehyde-induced catecholamine fluorescence is characterized microfluorometrically by correlation ratios of excitation maxima (370 : 320 nm and 320 : 275 nm).

A fluorometric method for simultaneous analyses of phenethylamine, phenylethanolamine, tyramine, and octopamine was developed by Kamata et al. (73). The method involves ion-exchange chromatography, derivatization with fluorescamine, solvent extraction, and then separation by thin-layer chromatography. The fluorescent spots are quantitated by scanning. The detection limits of this method are approximately 10 pmol for phenethylamine, phenylethanolamine, and tyramine, and 20 pmol for octopamine. The method was used for simultaneous analyses of putative neurotransmitter amines in whole rat brain.

Other ways of derivatization in combination with HPLC methods are described in Chapter 5, Section 1.

### 1.2. Amino Acids and Imino Acids

#### 1.2.1. Native Fluorescence

Aromatic amino acids show native fluorescence but differ considerably in their molecular extinction and fluorescence efficiency. Of the amino acids that occur naturally in proteins, only tryptophan and tyrosine (see Fig. 4.7) show appreciable fluorescence in aqueous solution (74). Data are summarized in Table 4.4.

The native fluorescence of some amino acids can be intensified by the addition of an organic solvent (e.g., ethanol, dimethyl sulfoxide) to

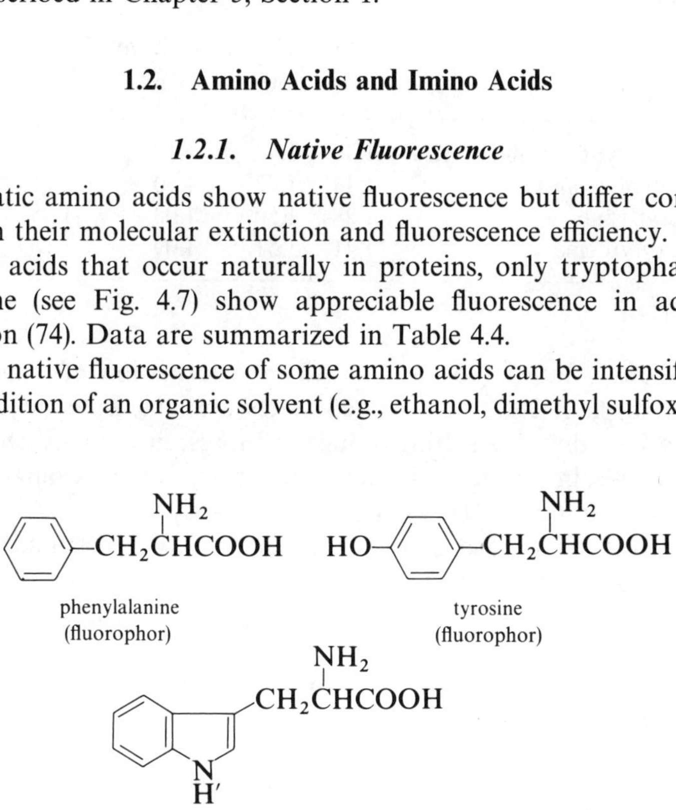

Figure 4.7. Structures of some aromatic amino acids.

### Table 4.4. Native Fluorescence Data for Some Aromatic Amino Acids (74)

| Amino Acid | Excitation Wavelength [nm] | Emission Wavelength [nm] | Fluorescence Efficiency [%] |
|---|---|---|---|
| Tryptophan | 287 | 348 | 20 |
| Tyrosine | 275 | 303 | 21 |
| Phenylalanine | 260 | 282 | 04 |

### Table 4.5. Changes in Fluorescence Intensities due to Addition of Organic Solvent (75) (Values are relative to fluorescence intensities in an aqueous solution)

| | Relative Fluorescence Intensity | | | | | |
|---|---|---|---|---|---|---|
| | [% Ethanol] | | | [% Dimenthylsulfoxide] | | |
| Amino Acid | 10 | 25 | 50 | 10 | 25 | 50 |
| 5-Hydroxy-tryptophan | 1.21 | 1.30 | 1.49 | 1.17 | 1.31 | 1.42 |
| p-aminophenylalanine | 1.22 | 1.48 | 1.72 | 1.50 | 2.11 | 2.87 |
| 3-Aminotyrosine | 1.20 | 1.28 | 1.50 | 1.08 | 1.25 | 2.00 |
| 3-Methoxytyrosine | 1.23 | 1.41 | 1.51 | 1.09 | 1.16 | 1.22 |

the aqueous solution (75). In this way the formation of complexes between the molecules in excited states and the water molecules seems to be reduced, resulting in higher fluorescence intensities. This effect depends, however, on the functional groups at the aromatic ring and on the aliphatic rests (see Table 4.5).

Fluorescence methods are widely used to study the changes in the environment of aromatic side chains within protein molecules. The fluorescence spectra of proteins in the presence of neutral salts were studied by Altekar (76). Generally, anions quench the protein fluorescence and can be arranged in the following sequence for their effectiveness as quenchers:

$$SO_4^{2-} \cong CH_3COO^- \cong Cl^- < ClO_4^- < SCN^- < Br^- < I^- < NO_3^- < NO_2^-$$

For analytical determinations, amino acids are usually derivatized with o-phthalaldehyde or other reagents, some of which were described in Section 1.1. Some selected applications are presented in the following section.

### 1.2.2. Derivatization

**1.2.2.1. Compounds With *o*-Phthalaldehyde.** A common method for the fluorometric determination of amino acids is derivatization with *o*-phthalaldehyde in combination with a thiol compound (see Section 1.1).

This reaction was studied with the fluorescence stopped-flow technique by Trepman and Chen (77). The results were consistent with the reaction of the amino compound with a 1 : 1 adduct of *o*-phthalaldehyde and the thiol compound. The equilibrium constant for the formation of the adduct from *o*-phthalaldehyde and mercaptoethanol was 164 $mol^{-1} dm^3$. A survey of the rates of reaction of this adduct with various amino acids demonstrates that with *o*-phthalaldehyde-mercaptoethanol-amino compound equal to 1 : 2.4 : 1, the reaction follows second-order kinetics, with $k = 150-450$ $mol^{-1} dm^3 s^{-1}$ at pH 9.0. The differences in rates are discussed in relation to structural differences between the amino acids. The rate of reaction of the *o*-phthalaldehyde/mercaptoethanol adduct with alanine is maximal at pH 10.5–11, and a great excess of mercaptoethanol results in a slower rate. Slower rates are also observed if dithiothreitol or 1-propanethiol is applied instead of mercaptoethanol.

During the derivatization reaction with *o*-phthalaldehyde, interference can be caused by glassware contamination (78). $NH_4^+$ ions give significant fluorescence with a spectrum similar to that of albumin and glycine, and further interference may be produced by crertain wash liquid. Thus extreme care is required when preparing glassware for use with this derivatization reaction.

In the following, some applications of the *o*-phthalaldehyde derivatization are summarized.

The detection of amino acids by automated ion-exchange chromatography using *o*-phthalaldehyde as the fluorogenic reagent is discussed by Thomas (79). Data are given for the fluorophor development, the relative fluorescent yields of various amino and imino acids, and the reproducibility of the method.

The fluorometric determination of amino acids dissolved in seawater by means of derivatization with *o*-phthalaldehyde is described by Sugimura and Suzuki (80). Free amino acids are reacted directly with *o*-phthalaldehyde, whereas total amino acids are pretreated by acid hydrolysis. The reaction product is separated from the solution by adsorption on XAD-2 resin and eluted from the resin by methanol. The recovery was 93.9% for total amino acids.

An improved method for the fluorometric analysis of amino acids is described by Yokotsuka and Kushida (81). This method includes the oxidation of proline and hydroxyproline with NaClO after ion-exchange fractionating and fluorophor formation with o-phthalaldehyde in the presence of 2-mercaptoethanol and a high concentration of Brij 35. Fluorescence intensities of proline and hydroxyproline are similar to the intensity of alanine, and fairly good reliability in the picomole range is achieved.

An automated cation-exchange fractionation and fluorometric detection of several cycle imino acids of plant origin is reported by Bleecker and Romeo (82). cis-4-Hydroxyproline, trans-4- and cis- and trans-5-hydroxypipecolic acids, acetylaminopipecolic acid, and 2,4-trans-4,5-trans- and 2,4-trans-4,5-cis-4,5-dihydroxypipecolic acids are resolved with a pH-3.1 buffer, and pipecolic acid is subsequently eluted by increasing the pH to 3.5. Detection involves a postcolumn oxidation of the secondary amines by chloramine-T before the introduction of the fluorogenic agent o-phthalaldehyde. The method establishes the usefulness of o-phthalaldehyde for the detection of nonprotein amino acids in plant material.

A two-column method for the rapid manual determination of glutamic acid, glutamine, GABA, and aspartic acid from brain regions is presented by Kimes and Shellenberger (83). The method uses a Dowex-1 column and various concentrations of HOAc to separate glutamate, aspartate, and the neutral amino acids. The neutral amino acids are subsequently placed on a Dowex-50 column, from which KOAc buffers are used to separate the amido amino acids are hydrolyzed in NaOH and the glutamine-derived glutamate is separated from contaminants on another Dowex-1 column. The amino acids are manually quantitated by using the fluorescence from the reaction with o-phthalaldehyde.

The procedure was designed to determine levels of amino acid neurotransmitters and catecholamines from a single brain sample as small as 0.12 g; however, because of the sensitivity of the o-phthalaldehyde reaction, levels of the four amino acids could be determined on a sample ≤ 10% of that size.

o-Phthalaldehyde is also applied as reagent for the analysis of amino acids in picomole quantities of peptides (84). For this highly sensitive method, an automated fluorescence amino acid analyzer is used. All commonly occurring amino acids, including cysteine, proline, and tryptophan, can be quantitated. High sensitivity is achieved primarily by using simple procedures that effectively and reliably reduce the level of interfering contamination present in buffers and reagents or introduced during sample handling. Accurate and reproducible results are obtained with ≤ 50 pmol peptide.

$o$-Phthalaldehyde or fluorescamine can be used as a fluorogenic reagent for the end-group determination of peptide (85). The method is based on the property that the derivatives of the $N$-terminal amino group of peptides formed in solution after reaction with either reagent are resistant to acid hydrolysis. The $N$-terminal amino acid can be determined by simply comparing the amino acid analysis of the original peptide with the fluorescent derivatives of the peptide. The decrease of the $N$-terminal residue in reacted peptides is 80–90% with fluorescamine and more than 90% with $o$-phthalaldehyde. Any $N$-terminal amino acid, with the exception of proline, can thus be detected.

**1.2.2.2. Compounds With Other Fluorogenic Reagents.** In addition to the derivatization with $o$-phthalaldehyde, several other fluorogenic reagents are used for amino acid analysis. Most of the reagents that form fluorescing derivatives with primary and secondary amines can be applied to amino acids (e.g., dansyl chloride, bansyl chloride, and NBD-chloride; (see Table 4.2).

The fluorometric determination of 16 amino acids at pH 4–9 using 8-hydroxyquinoline as fluorogenic reagent is described by Iqbal et al. (86). The detection limits range from 10 to 100 $\mu$mol m$^{-3}$.

Many secondary amino acids can be determined by reaction with 7-fluoro-4-nitrobenzo-2-oxa-1,3-diazole (NBD-F) at pH 7.5, at 70°C for 5 min and subsequent acidification to pH 1 (87). The detection limits for proline, hydroxyproline, and sarcosine are 0.08, 0.04, and 0.17 nmol cm$^{-3}$, respectively. Under the same conditions, the primary amino acids alanine, arginine, and aspartic acid are detected at 1.7, 1.7, and 3.4 nmol cm$^{-3}$, respectively.

The reagent 5-isothiocyanato-1,3-dioxo-2-($p$-tolyl)-2,3-dihydro-1$H$-benz(de)isoquinoline, described in Section 1.1, can also be used for the derivatization of amino acids (88). By reaction at 90°C, fluorescent thiocarbamyl derivatives are formed (analogously to the reaction in Fig. 4.2$g$), which have excitation maxima at 410 nm and emission maxima at 476 nm. The same reagent has also been applied for the derivatization of amino acids before separation by thin-layer chromatography (see below).

The effects of various surfactant micellar systems on the fluorometric determination of amino acids were studied by Singh and Hinze

(89). In the presence of cationic hexadecyl- or dodecyltrimethylammonium chloride and zwitterionic $N$-dodecyl-$N,N$-dimethylammonium-3-propane-1-sulfonic acid micelles, the fluorescence intensity of the derivatization product of glycine with dansyl chloride is enhanced. The lysine derivative of 2-mercaptoethanol-$o$-phthalaldehyde exhibits intensified fluorescence in the presence of nonionic Brij 35 or TX 100 and anionic Na dodecyl sulfate. Fluorescence enhancements of 8–20 were observed compared to the values obtained with aqueous solutions of the derivatized amino acids.

**1.2.2.3. Thin-Layer Chromatography.** A thin-layer chromatographic (TLC) method for the quantitative determination of amino acids is presented by Snejdarkova and Otto (90). After adjusting the pH to 10 with $NaHCO_3$ buffer and adding threefold excess glycine (as a carrier amino acid), and after reaction with a fivefold excess of dansyl chloride at 37°C for 30 min, two-dimensional chromatography on silufol is applied. The spots are eluted with ethanol/ammonium hydroxide, and fluorescence is measured at 505 nm (excitation wavelength, 342 nm).

A new approach for free amino acid analysis in plants is reported by Barcelon et al. (91). The free amino acids in the foliage and phloem exudate of several varieties of palm trees were separated with two-dimensional TLC on polyamide sheets and detected as derivatives of 5-dimethylaminonaphthalene-1-sulfonyl chloride (dansyl chloride; see Section 1.1).

Another TLC separation and detection of amino acids is described by Khalaf et al. (92). The amino acids are reacted with 5-isothiocyanato-1,3-dioxo-2-($p$-tolyl)-2,3-dihydro-1$H$-benz(de)isoquinoline (see also Fig. 4.2$g$) in alkaline or acid medium to form fluorescent thiocarbamyl or thiohydantoin derivatives. The products are extracted into $CHCl_3$ and chromatographed on micropolyamine plates.

Two sensitive methods for the detection and determination of phenylthiohydantoin amino acids by TLC are reported by Murayama and Kinoshita (93–95).

1. The phenylthiohydantoin amino acids were developed on a Kieselgel and $N$-chlorinated on the plate by spraying NaClO. The excess hypochlorite was removed by aeration, and the plate was sprayed with alkaline thiamin reagent. The chlorinated thiohydantoin gave an intensely fluorescent spot.

2. About 40–60 pmol of phenylthiohydantoin amino acids were detected on TLC plates by spraying $N$-chloro-5-dimethylaminonaphthalene-1-sulfonamide (NCDA). This reagent, which is pre-

pared by the action of sodium hypochlorite on dansylamide, is nonfluorescent but develops intense fluorescence with phenylthiohydantoin amino acids.

## 1.3. Alkaloids

Alkaloids are natural substances that usually occur in plants and have structures with at least one (generally heterocyclic) nitrogen atom (see Fig. 4.8).

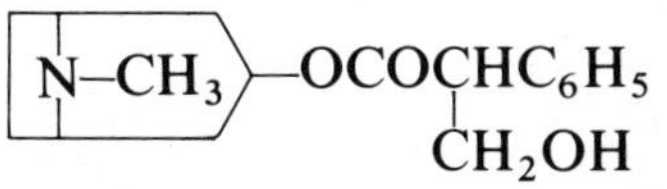

atropine

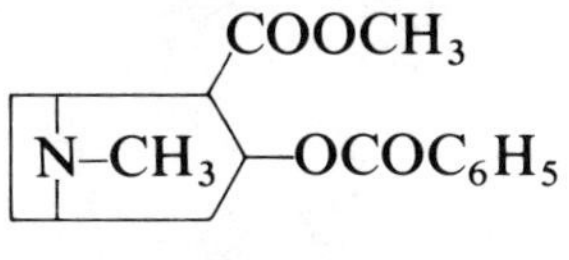

cocaine

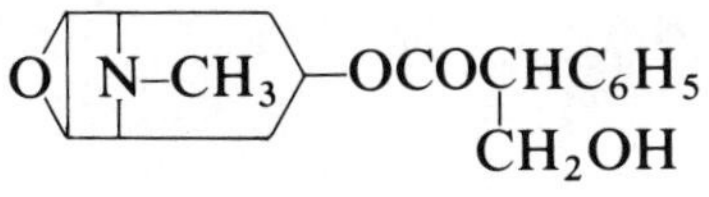

hyoscine
(scopolamine)

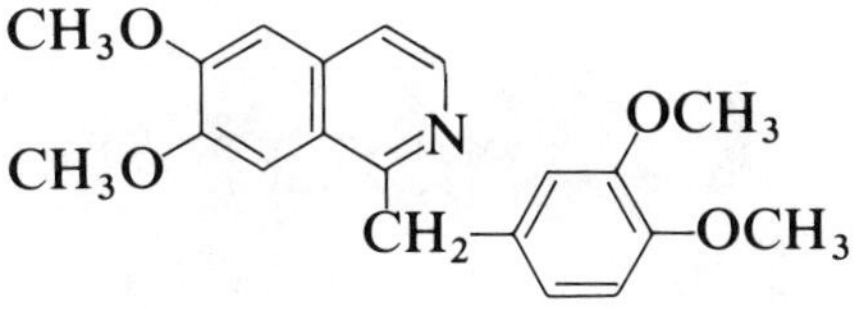

papaverine

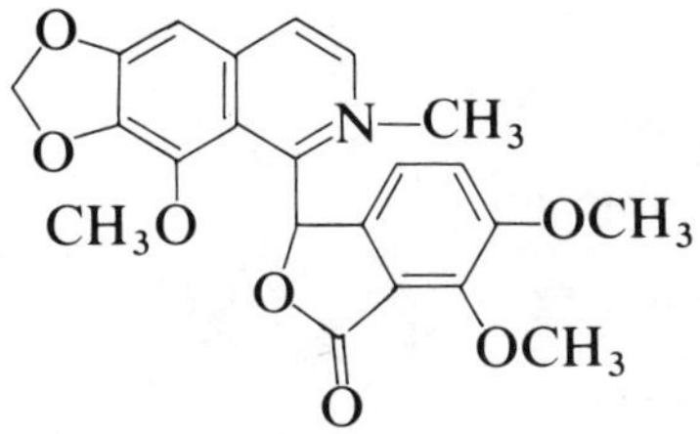

narcotine

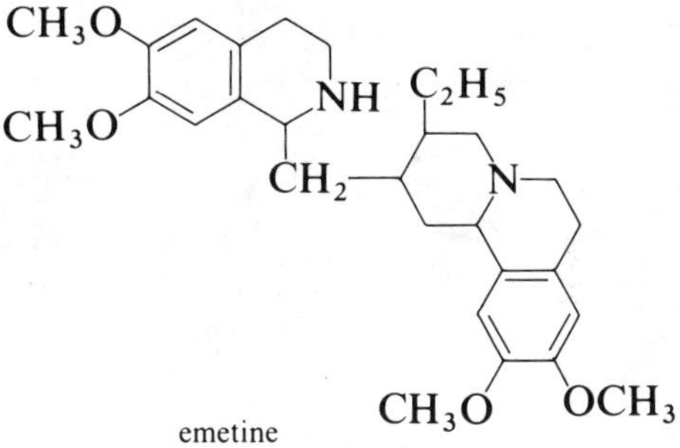

emetine

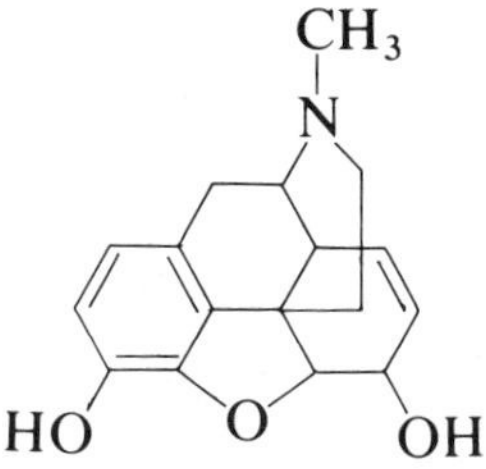


+ $H_2O$

merphine

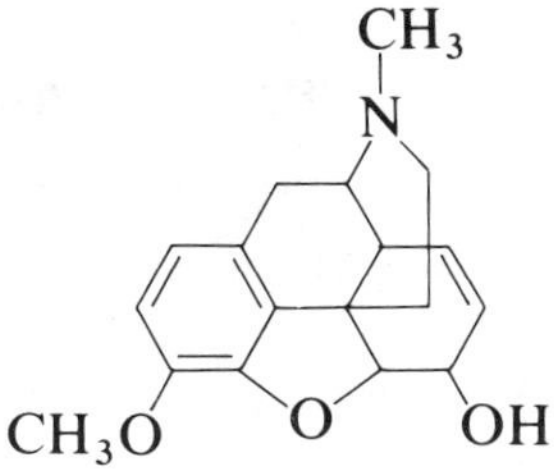


+ $H_2O$

codeine

**Figure 4.8.** Structures of some alkaloids.

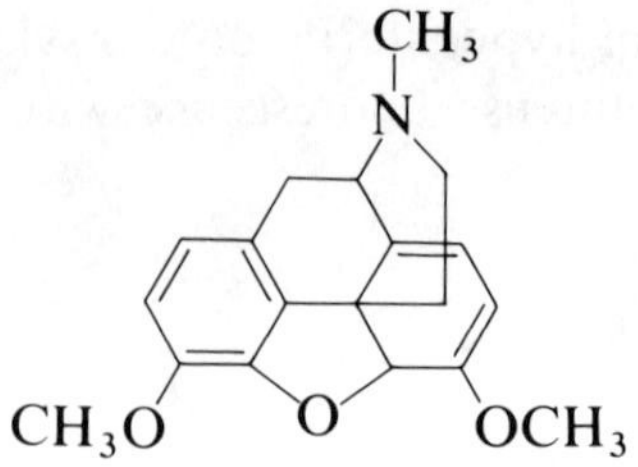


thebaine

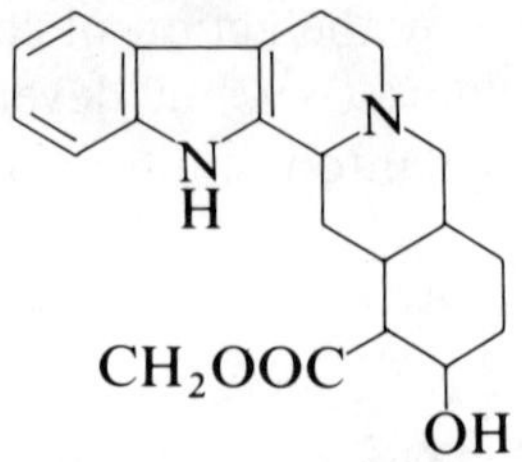


yohimbine

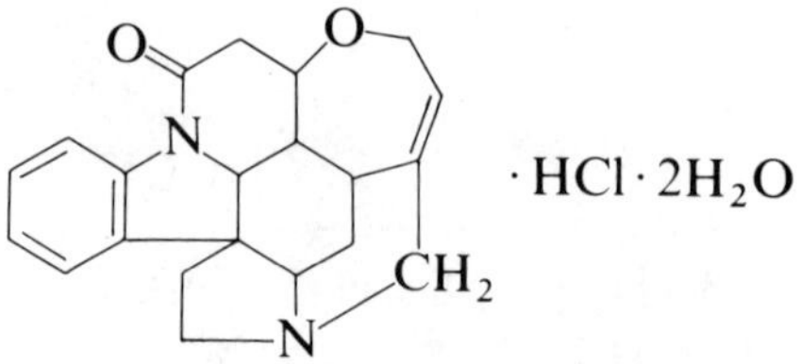


strychnine hydrochloride

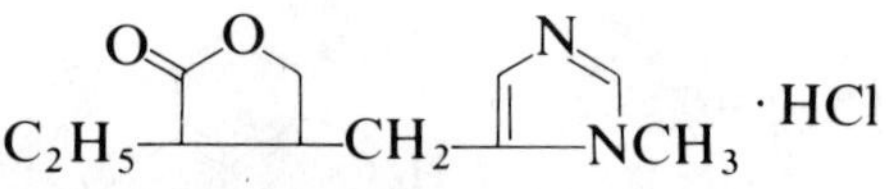


pilocarpine
hydrochloride

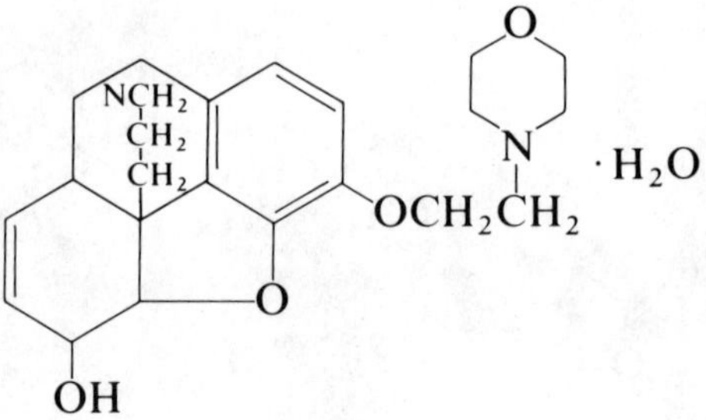


pholcodine

**Figure 4.8.** (*Continued*).

### *1.3.1. Native Fluorescence*

Several alkaloids can be determined in aqueous solution or in organic solvents by means of their native fluorescence. Some data are summarized in Tables 4.6 and 4.7.

Gürkan (99) reports fluorescence data for the alkaloids serpentine, yohimbine, and boldine in different solvents and in dependence of the pH values. At optimum conditions, detection limits are 0.5–5.0 ng $cm^{-3}$ (see Table 4.7).

**Table 4.6. Data Concerning Native Fluorescence of Some Alkaloids**

| Alkaloid | Wavelengths [nm] | | Reference |
|---|---|---|---|
| | Excitation | Emission | |
| Reserpine | 300 | 375 | 96 |
| Rescinnamine | 310 | 400 | 96 |
| Quinine | 250 | 350/450 | 96 |
| Papaverine | 320 | 348 | 97 |
| Narcotine | 335 | 400 | 97 |
| Balfourodinium salts | 310 | 345 | 98 |
| Platydesminium salts | 310 | 448 | 98 |

The quantitative analysis of the digitalis alkaloids digoxine and digitoxine in concentrated sulfuric acid after extraction is described by Naik et al. (100). Detection limits are in microgram range.

A fluorometric method for the determination of *balfourodinium* and *platydesminium* salts in *Choisya ternata* cell and tissue cultures is described by Montagu-Bourin et al. (98). These alkaloids were separated as ion-pair alkaloid-bromothymol blue complexes by TLC on silica gel G plates with EtOAc–HCOOH–$H_2O$ (10 : 1 : 1) as the mobile phase and determined by means of their native fluorescence, the data of which are given in Table 4.6. A method for the rapid estimation of *platydesminium* in total extracts is also presented.

The absorption and fluorescence properties of those two and of four other furoquinoline or pyranoquinoline alkaloids, extracted from Rutaceae plants, are also described by Montagu et al. (103). *Platydesminium* and *balfourodinium* were determined fluorometrically in *Choisya ternata* cell cultures and original plant extracts. For a powder plant sample, 95% recovery of the alkaloid was found.

A fluorometric determination of ipecacuanha alkaloids in pharmaceutical preparations is reported by Hassan (101). Ipecac tinctures and extracts are diluted with 0.05 $mol\,dm^{-3}$ $H_2SO_4$, and roots are triturated with DMSO, shaken with EtOH-0.005 $mol\,dm^{-3}$ $H_2SO_4$ (1 : 3), filtered, and diluted with 0.05 $mol\,dm^{-3}$ $H_2SO_4$.

The solutions are analyzed by fluorometry (excitation, 283 nm, emission, 318 nm). The major alkaloids, emetine and cephaline, have additive fluorescent intensities with maxima at pH 1–6, which are linear with concentration for 0.4–2.0 $\mu g\,cm^{-3}$. Recoveries were 98.98%, and relative standard deviations were < 0.01%.

Table 4.7. Some Applications of Fluorometric Analysis of Alkaloids

| Alkaloid | Solution | Wavelength (nm) | | Detection Limits (ng cm$^{-3}$) | Reference |
|---|---|---|---|---|---|
| | | Excitation | Emission | | |
| Serpentine | 0.05 mol dm$^{-3}$ H$_2$SO$_4$ | 304, 365 | 445 | 0.5 | 99 |
| Yohimbine | pH 10 | 285 | 370 | 5 | 99 |
| Boldine | 0.05 mol dm$^{-3}$ H$_2$SO$_4$ | 308 | 375 | 3 | 99 |
| | pH 10 | 330 | 420 | 1 | 99 |
| | Ethanol | 312 | 350 | 2 | 99 |
| Digoxin | H$_2$SO$_4$ (conc.) | 390 | 420 | | 100 |
| Digitoxin | H$_2$SO$_4$ (conc.) | 418 | 435 | | 100 |
| Emetine | 0.05 mol dm$^{-3}$ H$_2$SO$_4$ | 283 | 318 | | 101 |
| Cephaeline | 0.05 mol dm$^{-3}$ H$_2$SO$_4$ | 283 | 318 | | 101 |
| Pilocarpine | | 395 | 450 | 2.8 | 102 |
| Pholcodine | Derivatization with malonic acid and acetic acid anhydride | 395 | 450 | 0.5 | 102 |
| Cocaine | | 395 | 475 | 0.1 | 102 |
| Strychnine | | 420 | 450 | 0.5 | 102 |

Fluorescence spectra of solutions of atropine, cocaine, hyoscine, and some related drugs were examined by Hetherington et al. (104). The emission parameters of the excited states involved are discussed in relation to the structure and configuration of the tropines. The data obtained provide a basis for the microanalytical determination of these drugs.

The fluorometric analysis of opium alkaloids is described by Chalmers and Wadds (97). After extraction of the alkaloids with $CHCl_3$ at pH 9, papaverine and narcotine show different spectral maxima (320/348 nm and 335/400 nm, respectively). When trichloroacetic acid is added, the maxima are shifted to higher wavelengths, and the fluorescence intensity of papaverine is reduced, whereas that of narcotine increases. Morphine and codeine are determined in the aqueous phase.

The investigation of morphine and related alkaloids by fluorescence is described by McLeod and West (105). A microflow cell containing a transparent gold micromesh electrode was designed for *in situ* fluorescence monitoring of electrogenerated species by frontal illumination. As a model fluorogenic reaction, the oxidative dimerization of morphine to the fluorescent pseudomorphine was studied (see Fig. 4.9). The fluorescence calibration graph was linear over the concentration range $1 \times 10^{-3}$ to $1 \times 10^{-6}$ mol dm$^{-3}$, and the limit of detection was 5 $\times 10^{-7}$ mol dm$^{-3}$. The procedure, which is selective and free from interference from most of the opium alkaloids, made it possible to assay morphine directly in *Papaveretum* tablets and *Omnopon* injection ampules.

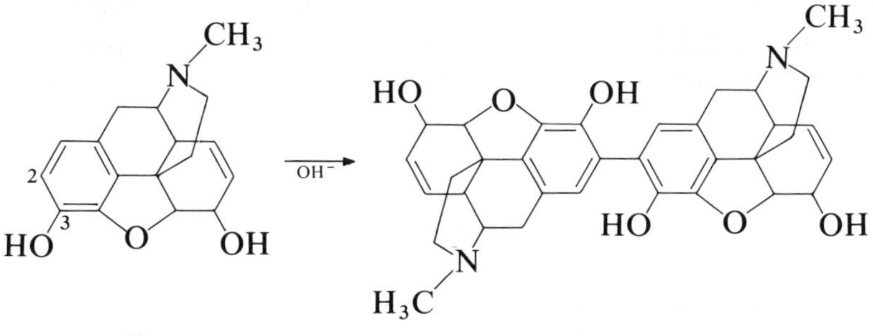


morphine
(nonfluorophor)

pseudomorphine
(fluorophor)

**Figure 4.9.** Oxidation dimerisation of morphine to the highly fluorescent pseudomorphine (2,2′-bimorphine).

### 1.3.2. Derivatization

Alkaloids with tertiary amino groups can be derivatized for fluorometric analysis by means of a mixture of malonic acid and acetic acid anhydride (102). The condensation requires 15 min at 80°C and shows high selectivity for this group of alkaloids. The detection limits are in the range of 0.1–2.8 ng cm$^{-3}$ (see Table 4.7).

Rao and Tandon (106) describe the application of malonic acid/acetic acid anhydride reagent for the spectrofluorometric determination of the opium alkaloids morphine, narcotine, codeine, papaverine, and thebaine. The alkaloid solution is heated with the anhydride reagent at 80–85°C for 20 min. After cooling and the addition of EtOH, the fluorescence is measured. The fluorescence intensity remains constant for 2 h for papaverine and $\leq$ 3 h in the case of the other alkaloids. The limits of detection of alkaloid are 0.5–4.0 ng cm$^{-3}$. However, there are several interferences with this determination, caused by tertiary amines, glucose, sodium chloride, potassium hydrogen phosphate, potassium chloride, and magnesium acetate. Water and other hydroxylated solvents inhibit the reaction if present from the start.

A TLC method for the simultaneous detection and *in situ* quantitation of ergolenes and some corresponding ergolines is described by Szabo and Karacsony (107). The method is based on the *o*-phthalaldehyde-$H_2SO_4$ fluorogenic reaction. For TLC, DC-Alufolien silica gel 60 foil is used as support. After application of a 5-$\mu$L sample aliquot, the plates are developed with EtOAc–MeOH (65 : 35), dried in the dark for 5 min, and sprayed immediately with the *o*-phthalaldehyde-$H_2SO_4$ reagent. Spots formed are stable and can be scanned even after several hours. Quantitation was carried out with a spectrometer equipped with a scanner, on the basis of the surface integral of the curves obtained. After spraying with the reagent, the intensity of the ergoline spots exceeds that of the ergolenes. The calibration curve is linear for 10–200 ng, and the method permits the detection of 1 ng of ergoline.

## 1.4. Vitamins and Related Compounds

Many vitamins consist of aromatic systems with substituent groups that endow the molecule with fluorescence in the visible region, or can simply be derivatized to form such structure (Fig. 4.10). Therefore, vitamins were probably the first group of biologically important compounds to be assayed fluorometrically.

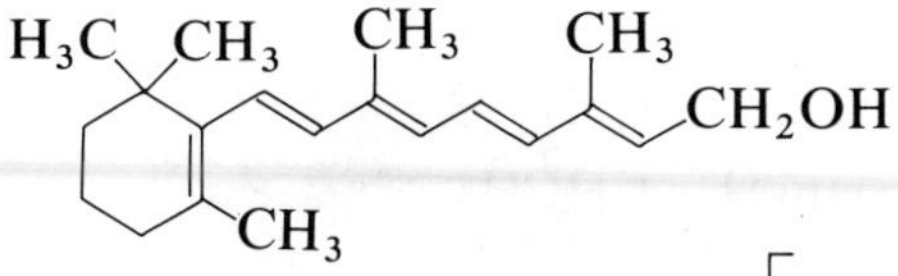

vitamin A

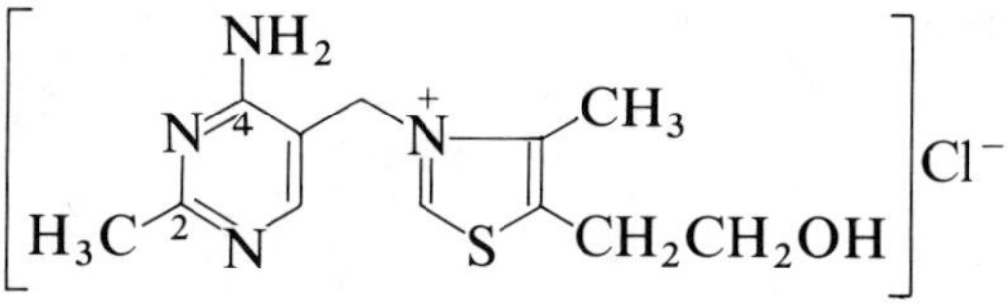

vitamin $B_1$ (thiamine)

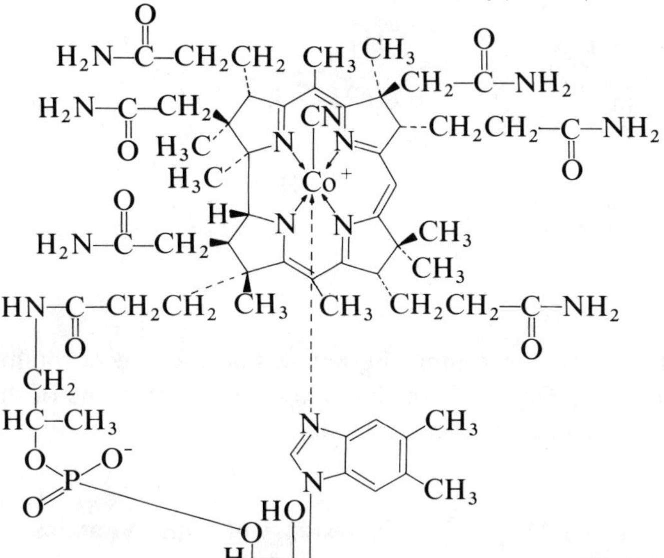
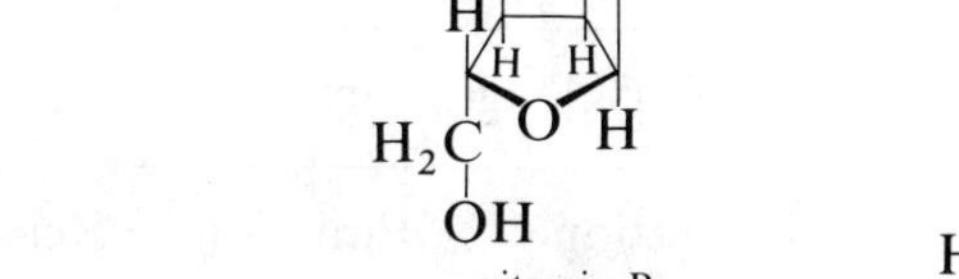

vitamin $B_{12}$

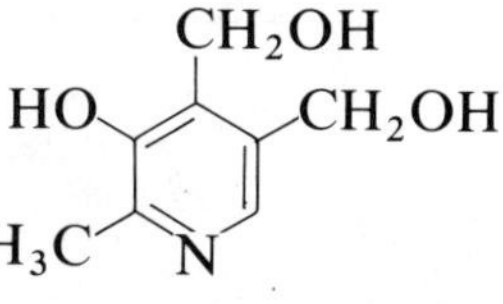

vitamin $B_6$ (pyridoxine)

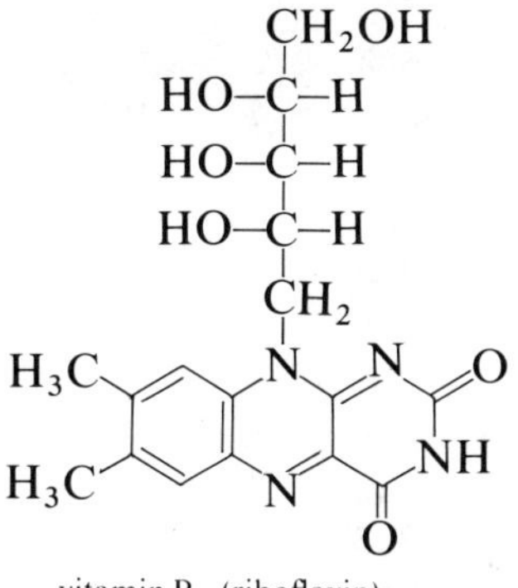

vitamin $B_2$ (riboflavin)

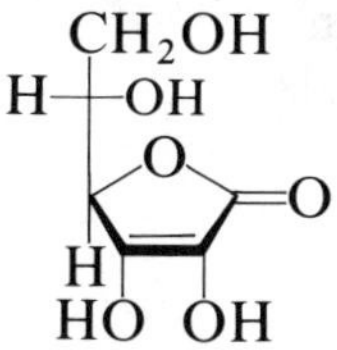

vitamin C
(ascorbic acid)

**Figure 4.10.** Structure formula of some vitamins.

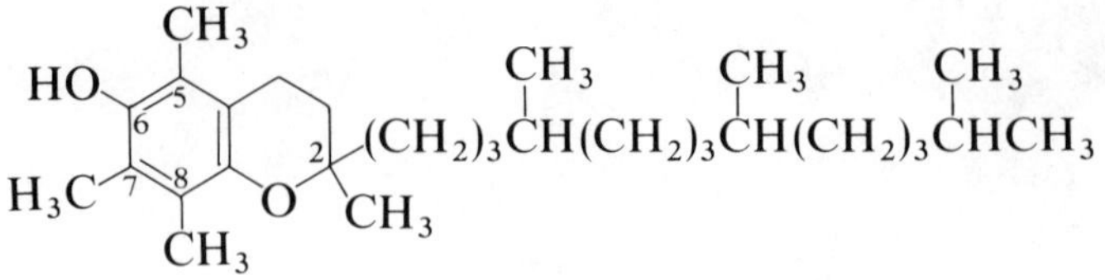

α-tocopherol

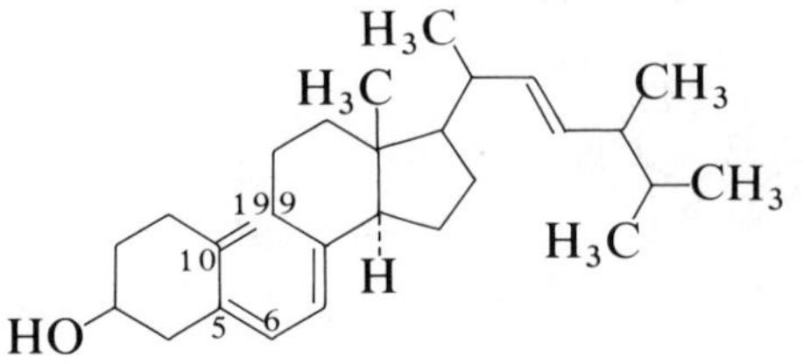

vitamin D₂

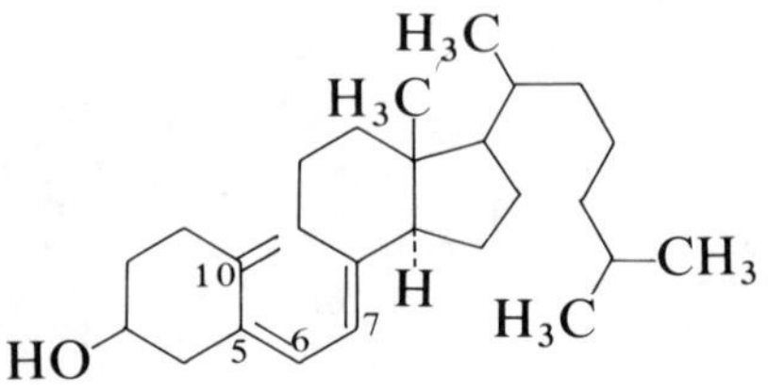

vitamin D₃

**Figure 4.10.** (*Continued*).

Spectral data concerning the native fluorescence of vitamins are summarized in Table 4.8. In the following section some of the compounds are discussed in more detail.

**Table 4.8. Native Fluorescence of Some Vitamins**

| Vitamin | Wavelength (nm) | | References |
|---|---|---|---|
| | Excitation | Emission | |
| A/abs. ethanol | 327 | 510 | 108 |
| A-acetate | 360 | 508 | 109 |
| $B_2$ (riboflavin) | 370 | 455–520 | 109 |
| $B_6$ compounds | | | |
|   Pyridoxine | 340 | 400 | 110 |
|   Pyridoxamine | 335 | 400 | 110 |
|   Pyridoxal | 330 | 385 | 110 |
| $B_{12}$ | 275 | 305 | 111 |
| C (ascorbic acid) | No native fluorescence | | |
| E (α-tocopherol) | 295 | 340 | 109 |
| p-Aminobenzoic acid | 294 | 345 | 111 |
| Nicotinamide | No native fluorescence | | |
| Folic acid | 365 | 450 | 111 |

### 1.4.1. Native Fluorescence

Vitamin A absorbs in the near-UV region because of the conjugated polyene system (for structures see Fig. 4.10). The detection limit for fluorometric measurements is about 1 ng cm$^{-3}$.

Kahan (112) reported a simple method for the quantitative assay of vitamin A in blood, using native fluorescence. Another fluorometric procedure for the determination of vitamin A in tissue homogenates and in fasting human plasma was published by Drujan et al. (113). After extraction of vitamin A with ethanol/diethyl ether from the blood sample, an aliquot of the solvent layer is evaporated and dissolved in chloroform/$n$-butanol.

Vitamin $B_2$ (riboflavin) can be determined using its native fluorescence (114, 115). Several procedures for the separation and fluorometric determination of riboflavin and related compounds in blood are summarized by Udenfriend (110).

Riboflavin and thiamin were determined in breakfast cereal, urine, and vitamin pills by Ryan and Ingle (116) using a multiple-wavelength fluorometer based on an intensified diode array detector. The native fluorescence of riboflavin in vitamin pills was monitored in one wavelength region, and the rate of formation in fluorescence thiochrome from thiamin (see Fig. 4.11) over another wavelength region, after a computer-controlled change in the pH of the reaction mixture.

Another method for the quantitative assay of riboflavin is its conversion to the more highly fluorescent lumiflavin (see Section 1.4.2).

Native fluorescence can also be used to identify and estimate the compounds pyridoxine, pyridoxamine, and pyridoxal, which belong

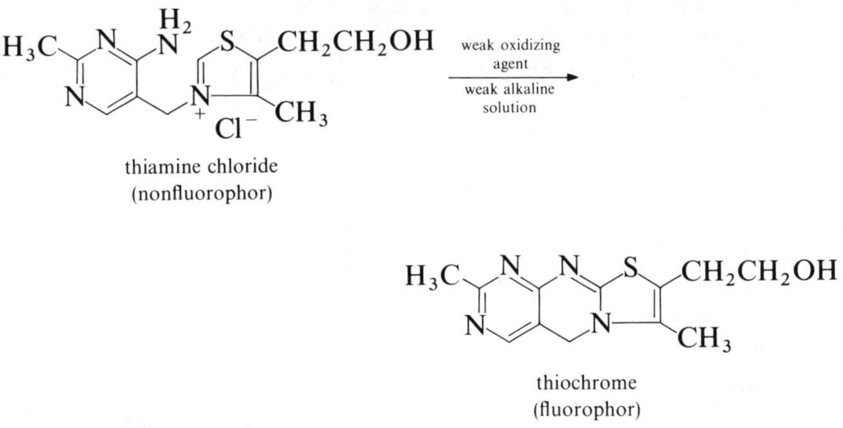

**Figure 4.11.** Conversion of vitamin $B_1$ (thiamine) to thiochrome.

to the vitamin $B_6$ group, in blood samples (117). The variation of fluorescence intensity and excitation and fluorescence wavelength with pH was studied in different dissociable forms of pyridoxamine by Bridges et al. (118). As reported by Chen (119), vitamin $B_6$ compounds show remarkable photosensitivity and variable temperature dependence. The fluorescence intensities of pyridoxamine and pyridoxamine phosphate change much more with temperature than the intensity of pyridoxal, so that problems occur in quantitative fluorometric studies of mixtures of vitamin $B_6$ derivatives.

Fluorescence characteristics and fluorometric methods for the assay of vitamin $B_6$ compounds are summarized in more detail by Udenfriend (110, 120).

Vitamin $B_{12}$ does not show the visible fluorescence that is characteristic of the porphyrin system, but exhibits only one fluorescent band in the far-UV region (111).

In comparison to chemical methods, the fluorometric determination of tocopherols (vitamin E) offers several advantages such as greater sensitivity and specificity and relatively mild conditions. A disadvantage, however, arises from the facts that $\alpha$-, $\beta$-, and $\gamma$-tocopherols show the same excitation and fluorescence spectra, but their molar fluorescence intensities vary with their molar extinction coefficients. For this reason precise analyses of mixtures of these compounds become complicated (110).

A procedure for the determination of tocopherols in blood and tissues using native fluorescence was developed by Duggan (121) and is also described in detail by Udenfriend (110).

The fluorescence intensity of *p*-aminobenzoic acid is rather high, but as this compound can easily be derivatized to form intensively colored products, chemical detection methods show sufficient sensitivity and are preferred in many applications. Fluorometric measurements of nicotinamide are possible only after derivatization (see Section 1.4.2).

### *1.4.2. Derivatizations*

Derivatizations are necessary when the vitamin itself is nonfluorescent (e.g., nicotinamide) or exhibits fluorescence of insufficient intensity (vitamin $B_1$). In some cases (riboflavin, folic acid), where also native fluorescence is used for quantitative determination, more sensitive measurements are possible after simple derivatization reactions.

Vitamin $B_1$ (thiamine) does not show appreciable fluorescence but can easily be oxidized to the fluorescent thiochrome (see Fig. 4.11) with excitation wavelength at 365 nm and fluorescence maximum at 430–435 nm [according to Ohnesorge and Rodgers (122) and Yagi et al. (123)]. This method is one of the standard procedures for thiamine assays in blood, tissue, urine, and other matrices. The determination of total thiamine in blood by this procedure was described by Burch et al. (124). As oxidation agent, an alkaline solution of potassium ferricyanide is used, and the thiochrome is extracted by $n$-hexyl alcohol after destroying excess oxidant with a reducing agent. An automatic system for the determination of thiamine using the thiochrome method was presented by Muiruri et al. (125) and by Pelletier and Madere (126).

The thiochrome method for determination of vitamin $B_1$ in pharmaceutical preparations was adapted by Karlberg and Thelander (127) to a continuous-flow system based on the flow-injection principle. A sample rate of $70 \, h^{-1}$ is easily attained if necessary. Results obtained with the system agree well with those obtained manually. The relative standard deviation is about 1%.

For fluorometric determination of nicotinamide, which itself is nonfluorescent, two steps of derivatization are necessary (see Fig. 4.12).

1. Methylation of nicotinamide with methyl iodide (128) or conversion of nicotinamide to a pyridinium product using cyanogen bromide solution (129).
2. Fluorometric assay as for $N$-methylnicotinamide by condensation with ketones such as acetone (130) or methyl ethyl ketone (131).

The nonfluorescent vitamin C (ascorbic acid) reacts with $o$-phenylenediamine to form a fluorescent quinoxaline derivative. The first application to the quantitative determination of ascorbic acid in solution was presented by Deutsch and Weeks (132). Reported wavelengths are 350 nm (excitation maximum) and 430 nm (emission maximum). The method is highly sensitive and specific in the presence of many other vitamins.

Vitamin $D_2$ and $D_3$ yield fluorescent products when treated with various strong acids (e.g., trichloroacetic acid), with excitation maximum at 390 nm and emission maxima at 470–480 nm (133, 134).

To increase the sensitivity of the fluorometric assay, riboflavin can be converted to lumiflavin, which is more highly fluorescent (see Fig. 4.13). This conversion, based on the cleavage between the ribitol unit and the flavin ring, is achieved by photodecomposition of ribo-

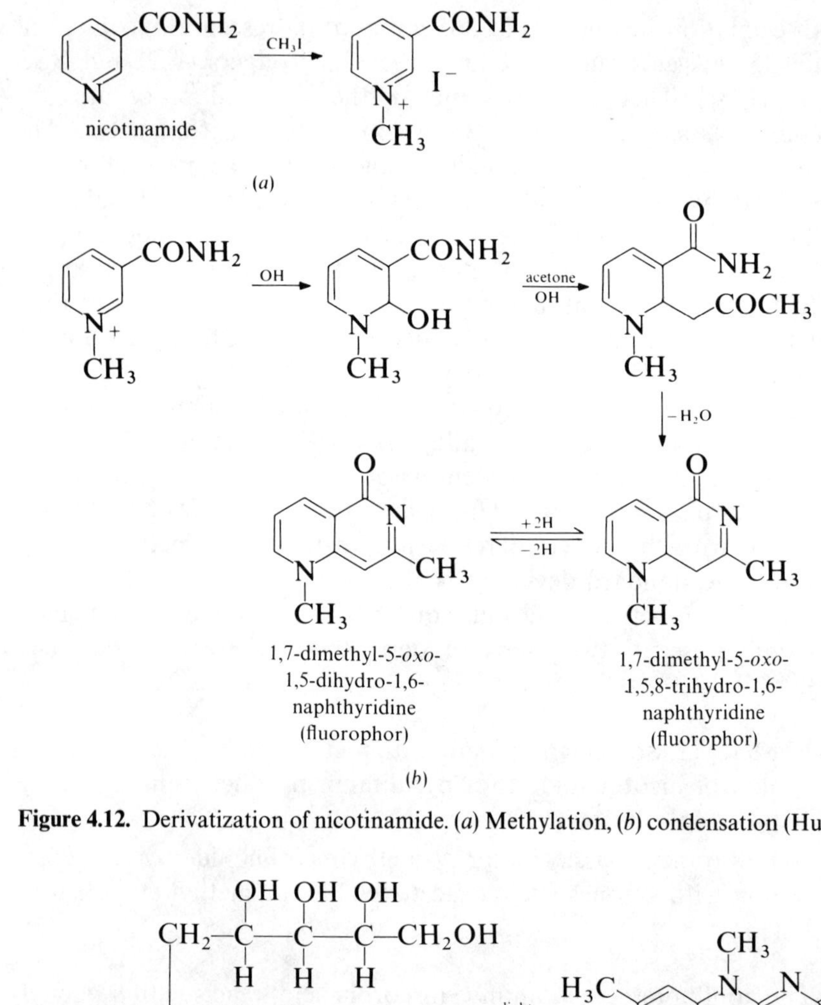

**Figure 4.12.** Derivatization of nicotinamide. (*a*) Methylation, (*b*) condensation (Huff).

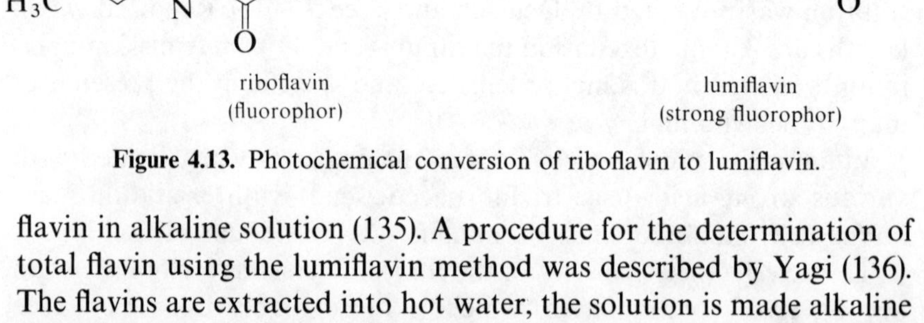

**Figure 4.13.** Photochemical conversion of riboflavin to lumiflavin.

flavin in alkaline solution (135). A procedure for the determination of total flavin using the lumiflavin method was described by Yagi (136). The flavins are extracted into hot water, the solution is made alkaline (pH 11), irradiated under a fluorescent lamp for about 30–60 min, and then acidified, and the lumiflavin is extracted into chloroform.

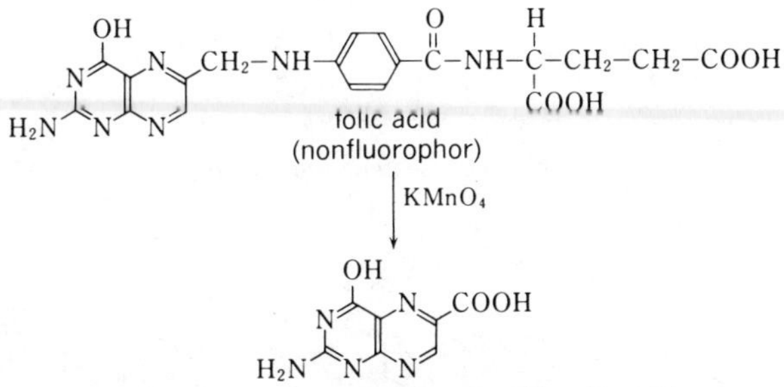


folic acid (nonfluorophor)

2-amino-4-hydroxypteridine-6-carboxylic acid (fluorophor)

**Figure 4.14.** Oxidation of folic acid with permanganate.

A method for the determination of folic acid (pteroylglutamic acid) in plant and animal tissues after derivatization is described by Allfrey et al. (137). Folic acid is split by oxidation with permanganate into a pterine and a nonpterine fragment. The pterine part is highly fluorescent 2-amino-4-hydroxypteridine-6-carboxylic acid (see Fig. 4.14). In the presence of interfering pigments, the fluorescent oxidation product is isolated chromatographically by adsorption on Florisil at pH 4 and elution in sodium tetraborate solution.

## 1.5. Steroids

Steroids are natural compounds consisting of a system of four condensed rings. In acid solution, almost all steroids show more or less intense flourescence, depending on the type and concentration of the acid, the temperature, and so on.

In this section some special applications for three major groups of steroids are presented.

### *1.5.1. Estrogens*

The structure of estrogens is different from that of the other steroids, for it includes an aromatic ring and thereby a phenolic OH group (see Fig. 4.15*a*). This phenolic structure has two effects that are important for analytical purposes.

1. Estrogens can be extracted from organic solutions into aqueous alkali and so be separated from other steroid compounds.
2. Estrogens are the only group of normally occurring steroids with the property of native fluorescence.

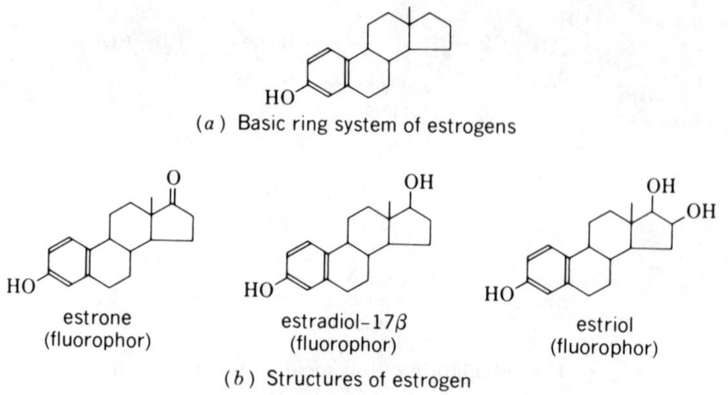

**Figure 4.15.** (a) Basic ring system of estrogens, and (b) structures of estrogens.

For some estrogens, data regarding native fluorescence are summarized in Table 4.9 (138). An application is described by Bramhall and Britten (139). Estrogens in urine samples can be determined by means of their native fluorescence after precipitation with ammonium sulfate and extraction. The practicable range of concentration is about 2 $\mu g\, cm^{-3}$.

When estrogens are heated with concentrated sulfuric acid, they show a yellow-green fluorescence. This reaction was reported by Kober (140) and applied for the fluorometric determination of estrogens. Sulfuric acid (85%) containing 2% quinone and 0.5% $Fe_2(SO_4)_3$ ("Kober's reagent") is heated to 95°C to produce fluorophors with estrogen compounds (138). The determination of estrogens by utilizing their fluorescence in sulfuric acid was modified by Ittrich (141, 142). The fluorophor is induced by the treatment of estrogens with less concentrated sulfuric acid in the presence of hydroquinone and is extracted into an organic solvent containing p-nitrophenol. When a

**Table 4.9. Native Fluorescence of Some Estrogens (138)**

| Estrogen | Wavelength (nm) | | Relative Intensity |
|---|---|---|---|
| | Excitation | Emission | |
| β-Estradiol | 285 | 330 | 100 |
| Estrone | 285 | 325 | 13 |
| Equilin | 290 | 345 | 10 |
| Equilenin | 250, 290, 340 | 370 | 1000 |

mixture of 1% ethanol and 2% *p*-nitrophenol in acetylene-tetrabromide is used, the maxima are about 540 nm (excitation) and 560 nm (emission). The Kober Ittrich reaction has become a standard method for the determination of estrogens in clinical chemistry (143–145).

The simultaneous determination of total estrogens by fluorometry and of pregnanediol and 17-ketosteroids by gas chromatography from the same urinary extract is described by Taylor and Carter (146). A 20-$cm^3$ urine sample is hydrolyzed with $\beta$-glucuronidase/aryl sulfatase, extracted with diethyl ether, and washed twice with 20 $cm^3$ solution of NaOH. While the organic phase is silylated and analyzed by gas chromatography, the NaOH extract is acidified with HCl, extracted with diethyl ether, and analyzed for estrogens by fluorometric detection.

To differentiate estrogen compounds, a chromatographic separation is performed before the fluorometric assay. Estrone, estradiol, and estriol can be separated by column chromatography using activated alumina (147). The benzene extracts of the samples are placed on the column, and the estrogens are eluted with benzene-methanol mixtures with increasing methanol concentrations in the sequence given above (for structures see Fig. 4.15*b*).

Separation of the three estrogens by paper chromatography using a glass-paper impregnated with *p*-toluenesulfonic acid is described by Epstein and Zak (148). Red to yellow fluorescence is achieved by heating the paper chromatograms and can be measured by scanning.

Estrone can be determined in the presence of estradiol and estriol when a 30% hydrogen peroxide is added to a mixture of these compounds in sulfuric acid. Under these conditions, only estrone gives blue fluorescence (149).

### 1.5.2. Cholesterol

The formation of fluorophors in sulfuric acid, which is applied for the determination of estrogens (see above), can also be utilized in cholesterol (Fig. 4.16) analysis. A fluorometric microprocedure is described

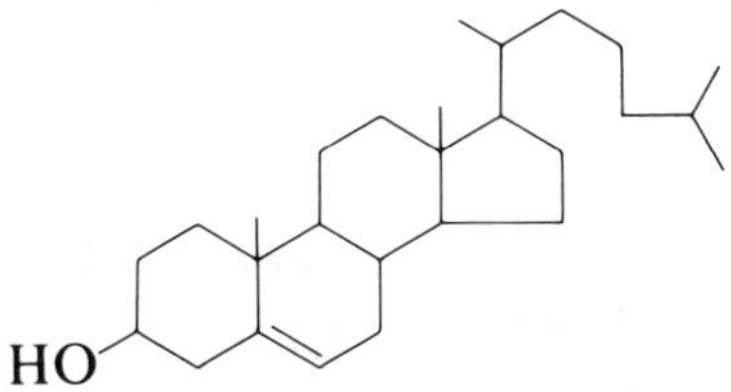

Figure 4.16. Structure of cholesterol.

by Albers and Lowry (150). After extraction from tissue samples with absolute ethanol and evaporation of the ethanol extract, the residue is dissolved in a mixture of trichloroethane and acetic anhydride (5 : 1) and mixed with concentrated sulfuric acid. Fluorescence is measured after 1–2 h. As the amount of cholesterol is several orders of magnitude higher than the amounts of other steroids, which are also extracted by this method, normally no serious problems are caused by interference (138).

A modification of this method and its application to small blood samples were presented by McDougal and Farmer (151). Four cubic millimeters of serum are added to 40 $mm^3$ of a solution of glacial acetic acid and trichloroethane (3 : 2) and mixed with 1 $cm^3$ of acetic anhydride/trichloroethane and 40 $mm^3$ of concentrated sulfuric acid. Fluorescence of the centrifuged solution is measured after 40 min.

For application to human plasma, the Albers–Lowry method (150) was modified by Koval (152). Samples of serum are precipitated with trichloroacetic acid, and cholesterol is extracted with a solution of potassium acetate in absolute alcohol. After addition of KOH, the sample is incubated at 40°C for 1 h. Petroleum ether is used to extract cholesterol, and after evaporation the residue is dissolved in a mixture of acetic anhydride–trichloroethane–concentrated sulfuric acid. The fluorescence is measured after about 30 min with maxima of about 546 nm (excitation) and 590 nm (emission).

Another method for the fluorometric determination of cholesterol in serum by addition of $FeCl_3$ is described by Solow and Freeman (153). There is no interference by pigments and ionic compounds, and the practical range of concentration for this method is about 10 ng $mm^{-3}$.

The automatized fluorometric analysis of cholesterol in serum is presented by Fruchart et al. (154). In the first step of reaction, cholesterol oxidase and esterase form hydrogen peroxide, which in the second step is condensed with formaldehyde and ammonium hydroxide to 3,5-diacetyl-1,4-dihydrolutidine. The spectral maxima are 405 nm (excitation) and 485 nm (emission).

### 1.5.3. *Other Steroids*

In strong sulfuric or phosphoric acid, adrenal cortical steroids (see Fig. 4.17) with a hydroxy group at position 11, such as corticosterone and hydroxycortisone (but not aldosterone), form fluorophors with excitation maxima at 470–475 nm and fluorescence maxima between 520 and 530 nm [Sweat (155), Kalant (156), Goldzieher and Besch (157)].

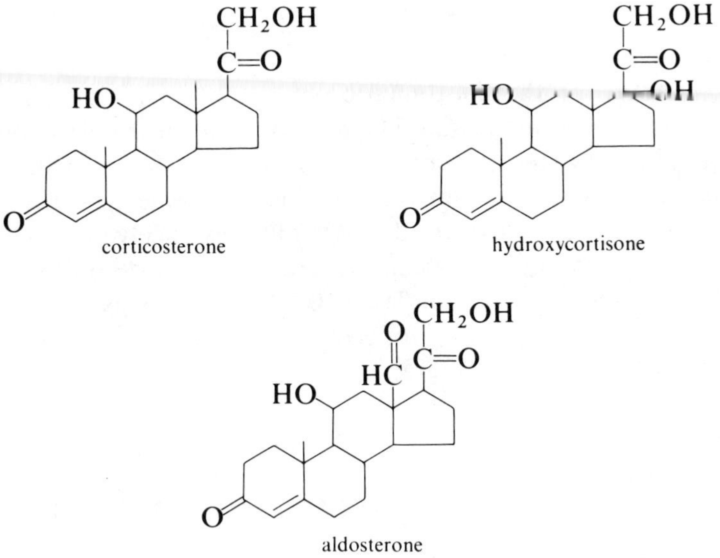

**Figure 4.17.** Structures of some adrenal steroids.

A method for the determination of total 11-hydroxyadrenal steroids is presented by Zenker and Bernstein (158) and Silber (159) and is also described in detail by Udenfriend (138).

The assay of plasma corticosteroids can be performed by the following method (160). To eliminate liquid materials, 4 cm$^3$ of plasma is extracted with 12 cm$^3$ of petroleum ether, and then mixed with 15 cm$^3$ of methylene chloride. After centrifugation and removal of the plasma layer by aspiration, the organic phase is extracted with 1 cm$^3$ of NaOH to remove estrogens. An aliquot of the methylene chloride is shaken with a mixture of concentrated sulfuric acid and ethanol (3 : 1) and centrifuged. The fluorescence of the acid layer is measured after 6 min (143).

Fluorescence of adrenal cortical steroids can also be induced in alkaline media. As reported by Bush and Sandber (161), $\Delta^4$-3-ketosteroids show an orange fluorescence on paper chromatograms when the paper is sprayed with a solution of sodium hydroxide and examined under ultraviolet light. Fluorescence can be measured directly on the paper (162, 163), with excitation maximum at about 365 nm and emission maxima between 550 and 600 nm for all $\Delta^4$-3-ketosteroids. The same sort of fluorescence is achieved in solution when the reaction is carried out in potassium *tert*-butoxide dissolved in *tert*-butyl alcohol (164). The spectral maxima are about 385 nm (excitation) and 580 nm (emission).

A method for the fluorometric determination of ecdysteroids is described by Koolman (165). The procedure is linear over two orders of magnitude of steroid concentration, sensitive down to $10^{-11}$ mol ecdysteroid, and specific for ecdysteroids. Ecdysone gives the highest fluorescence, other ecdysteroids having fluorescence values 50% relative to ecdysone. Interference in the assay by other steroids, including cholesterol and 7-dehydrocholesterol, is low. The assay is suitable for detection and quantification of ecdysteroids in extracts of biological materials after a chromatographic step.

## 2. BIOMEDICAL AND CLINICAL CHEMISTRY

### N. ICHINOSE and K. ADACHI

Biomedical and clinical samples pose a set of unique problems of their own because their matrices include blood, sera, urine, tissues, protein hydrolysates, various metabolites, and so on containing a large number of compounds in which the absorption or emission spectra of other, coexisting compounds present may overlap the spectrum of a desired component.

Recently, the fluorometric method as a novel analytical technique for such biomedical and clinical samples is becoming one of the most important applications of the photoluminescence phenomenon, which is very useful for medical examination and diagnosis since the fluorometric method offers several advantages over the conventional methods: high sensitivity, selectivity, and wide linear range. Up to now, then, many approaches have already been developed to enable analysts to overcome some of the difficulties, such as the above mentioned ones, associated with the use of photoluminescence with their samples.

In this section, we deal with current and significant applications of luminescence spectrometry with individually special relevance to problems in biomedical and clinical chemistry. Coverage is limited to publications that appear to place special emphasis on emerging theory and analytical techniques of broad interest in the various fields of biomedical and clinical chemistry during the past 10 years.

## 2.1. Biomedical Chemistry

### 2.1.1. Autoanalysis by Fluorometry of Enzyme Estimation of Free and Total Cholesterol (154)

A fluorometric, automatic method for determination of free and total cholesterol in serum is described by Fruchart et al. (154). Cholesterol esters are split into free cholestrol and fatty acids by cholesterol esterase. In the presence of oxygen, free cholesterol will be transformed by cholesterol oxidase into $\Delta^4$-cholestenone with the formation of hydrogen peroxide. Hydrogen peroxide oxidizes in the presence of catalase methanol to formaldehyde, which gives, with ammonium ions and acetylacetone, 3,5-diacetyl-1,4-dihydrolutidine (Fig. 4.18), measured by fluorimetry. Without cholesterol esterase, the reaction allows a measurement of free cholesterol in plasma. Therefore, the ratio of ester to total cholesterol is determined by the two steps. Accuracy and precision are very good. Results are well correlated with those obtained with reference methods.

### 2.1.2. Fluorometric Determination of Ammonia in Protein-Free Filtrates of Human Blood Plasma (166)

Various methods are available to determine the concentration of ammonia in whole blood or plasma. The most sensitive method for measuring ammonia is the fluorometric technique of Rubin and Knott (167). Although these authors recommend the use of untreated serum, Spooner et al. (166) have found this unsatisfactory in more than 50% of routine samples, largely because of quenching and turbidity problems. They have therefore applied fluorometric techniques to the determination of ammonia in protein-free supernates of plasma using a sequential technique based on monitoring NADH oxidation and a batch technique based on determination of $NAD^+$.

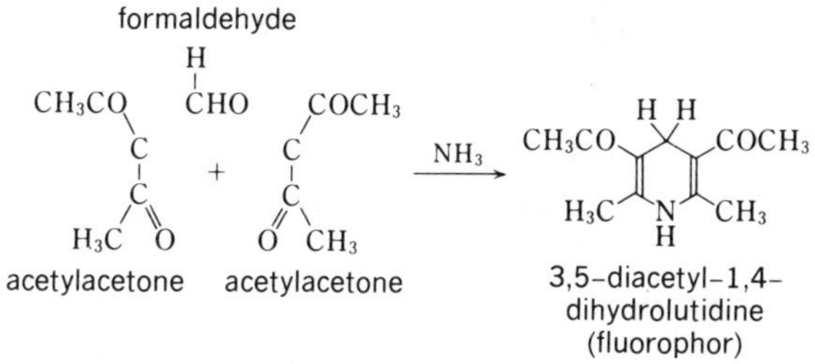

**Figure 4.18.** Formation of fluorophor from formaldehyde, produced through enzymatic decomposition, for the fluorescent determination of cholesterol.

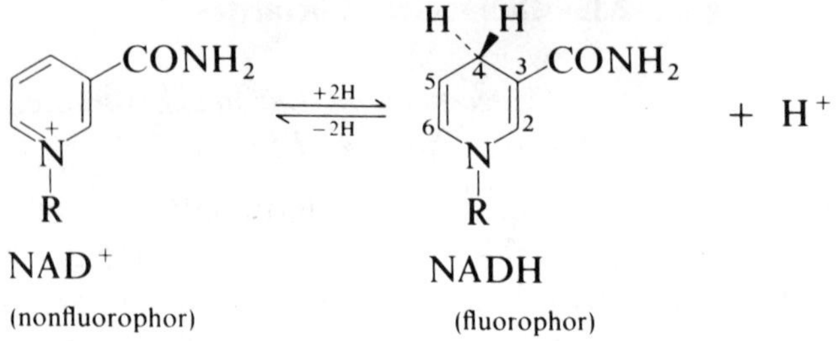

108    BIOCHEMICAL AND BIOMEDICAL APPLICATIONS

**Figure 4.19.** Structure of NADH.

Ammonia has been determined in filterates of human plasma after precipitation of the proteins by perchloric acid. After restroration of pH to around 7.5, addition of 2-oxoglutarate, NADH, and glutamate dehydrogenase (GDH) converts the ammonia to L-glutamate with oxidation of the NADH to NAD$^+$ [Eq. (4.1); Fig. 4.19]. This latter reaction was utilized in two ways. In the first reduction of native NADH fluorescence under the conditions of the GDH reaction provided a measure of ammonia concentration. In the second, residual NADH was destroyed by acid treatment, and the fluorescent product generated from NAD under strongly alkaline conditions was assayed. The optimal requirements for both methods were defined, their linearity and precision ascertained, and their relative merits compared. The first method was convenient for "one-off" estimations, and the second for larger batches.

Ammonia concentration increased in plasma and in acid protein-free filtrates of plasma irrespective of the conditions of storage; however, when the latter were neutralized, storage at $-20\,°C$ was effective.

$$NH_4^+ + NADH + 2\text{-oxoglutarate} \xrightleftharpoons{GDH} \text{L-glutamate} + NAD^+ \quad (4.1)$$

### 2.1.3. Fluorometric Determination of Tetracyclines in Biological Materials (168)

A method for the fluorometric determination of tetracyclines in biological materials based on solvent extraction of mixed tetracycline-calcium trichloroacetate ion pairs from aqueous solutions is described by Poiger and Schlatter (168). The extraction of tetracycline; (TC; Fig. 4.20) and chlorotetracycline (CTC) is almost quantitative, whereas only very poor extraction occurs with oxytetracycline (OTC). However, saturation of the aqueous phase with sodium chloride results in complete extraction of OTC into the organic phase. This

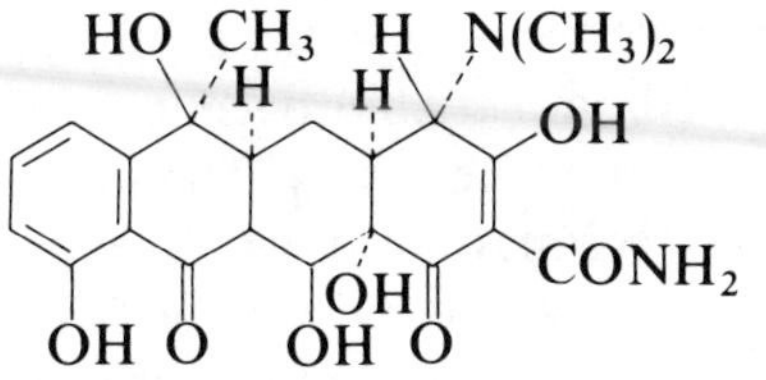

**Figure 4.20.** Structure of tetracycline.

effect enables OTC to be determined in the presence of TC and CTC. Ethyl acetate was found to be the most suitable extractant. The fluorescence of the organic phase is measured after addition of magnesium ions and a base.

The excitation maxima of all three tetracycline derivatives are about 400 nm and emission maxima are 500 nm. They differ only very slightly and cannot be used for differentiation between the derivatives.

### 2.1.4. Erythrocyte Uroporphyrinogen I Synthase Activity in Diagnosis of Acute Intermittent Porphyria (169)

Measurement of the activity of uroporphyrinogen I synthase provides an excellent laboratory aid in the diagnosis of acute intermittent porphyria, particularly in patients who are asymptomatic or in whom the disease is not biochemically manifested by porphyrin-precursor exeretion. The method described here is a simplified fluorometric procedure for measuring the activity of this enzyme in whole blood. The assay is based on a coupled-enzyme procedure in which added δ-aminolevulinic acid and the dehydratase that is present in erythrocytes are used to generate porphobilinogen (PBG) as substrate for

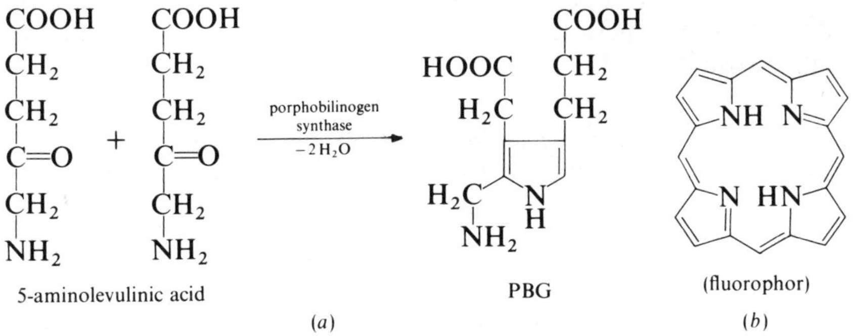

**Figure 4.21.** (a) PBG as a intermediate in biosynthesis of porphyrin, (b) porphyrin ring.

uroporphyrinogen synthase (Fig. 4.21). After appropriate incubation the protein is removed with trichloroacetic acid, and the porphyrins formed are measured fluorometrically. The sensitivity, specificity, and precision of the assay compare well with previously described procedures. Activity in nonporphyric male subjects was 31 (S.D., 6.0) nmol of porphyrin formed per milliliter of erythrocytes per hour at 37°C. Application of the method for identifying gene carriers of acute intermittent porphyria is demonstrated in three generations of an affected family.

### 2.1.5. Quantitative Analysis of Bile Acids and Their Conjugates in Duodenal Aspirate by Fluorometry after Cellulose Acetate Electrophoresis (170)

Since the clinical importance (171) of bile acids and their conjugates and metabolites is firmly established, a rapid, specific, and sensitive method for their determination is needed. Numerous methods for separation and quantitative determination of bile acids and salts have been investigated. Most of these techniques are based on paper (172), thin-layer (173) or column-adsorption chromatography (174), ion-exchange (175) or gas chromatography (176), and paper electrophoresis (177).

The present method combines electrophoretic separation on cellulose acetate strips and their quantitative enzymatic determination by 3α-hydroxysteroid dehydrogenase. This method allows an accurate and specific determination of the total bile salt content, the ratio of glycine to taurine conjugates, and the ratio of deoxycholic to cholic conjugates (Fig. 4.22).

Duodenal aspirates are directly applied to cellulose acetate without any previous purification. Bile salts are separated by electrophoresis into seven bands. After enzymatic reaction, quantitative fluorometric determination is performed on each band of nonsprayed cellulose acetate strips.

### 2.1.6. Enzymatic Spectrofluorometric Determination of Uric Acid in Microsamples of Plasma by Using p-Hydroxyphenylacetic Acid as a Fluorophor (178)

A sensitive spectrofluorometric micromethod for the determination of uric acid is presented together with its application to human and rat plasma by Sumi et al. (178). The method is based on the reactions of uricase and peroxidase coupled with p-hydroxyphenylacetic acid as a fluorophor (Fig. 4.23).

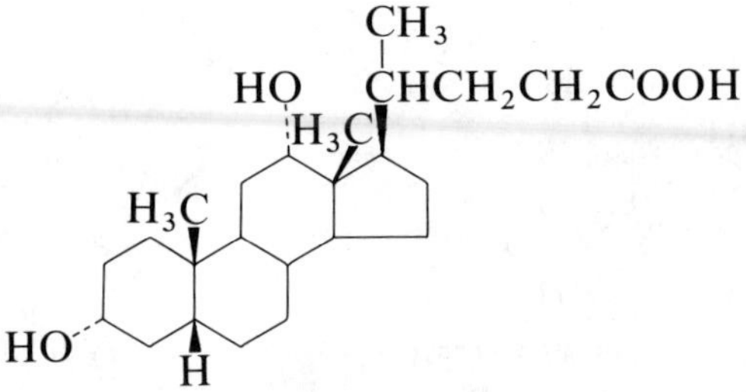


deoxycholic acid
(fluorophor)

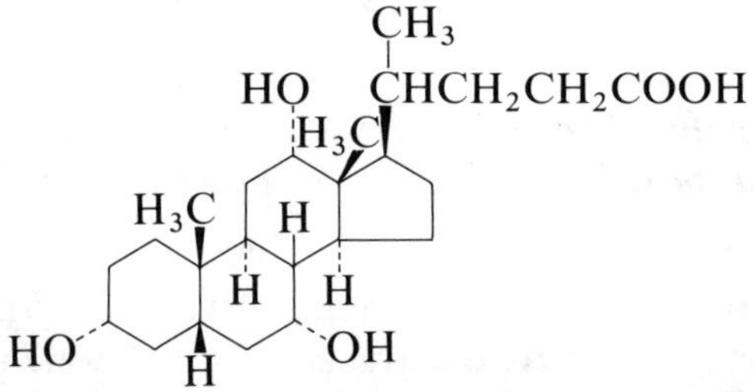


cholic acid
(fluorophor)

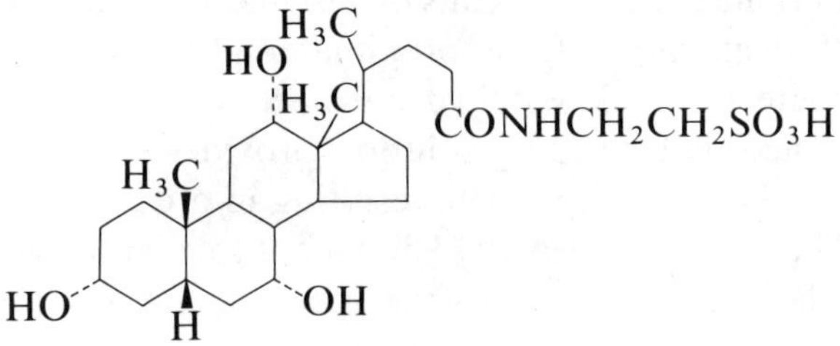


taurocholic acid
(fluorophor)

**Figure 4.22.** Bile acids and their conjugates.

There is a very wide range of proportionality between the concentration of uric acid and the increase of fluorescence intensity. Ten nanograms of uric acid is determinable. At most, 25 $mm^3$ of human or 50 $mm^3$ of rat plasma is sufficient to obtain the accurate value of endogenous plasma uric acid concentration.

To test the applicability of the method, analysis of picomole amounts was performed in the islets of Langerhans and in tissue samples of much smaller size than are used in fine-needle biopsies.

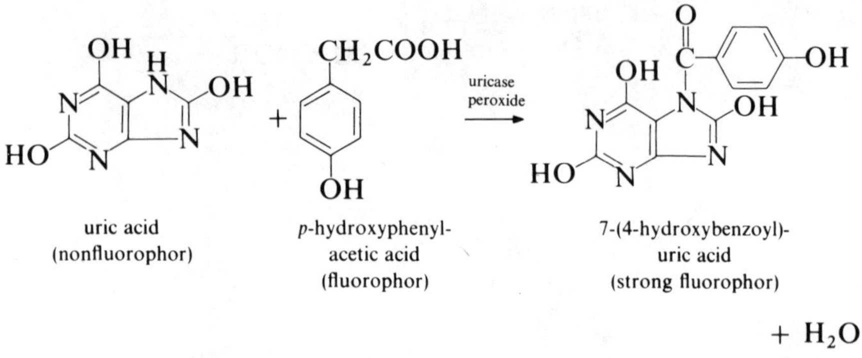

Figure 4.23. Fluorescent derivatization of uric acid.

### 2.1.7. Simplified Luciferase Assay of $NAD^+$ Applied to Microsamples from Liver, Kidney, and Pancreatic Isolets (179)

A new single-step procedure for the bioluminescence assay of $NAD^+$, permitting measurements on the picomole level ($10^{-12}$ mol), is described by Ågren et al. (179). Acid extracts of $NAD^+$ were prepared in different tissues. The acidification destroys reduced pyridine nucleotides and most enzymes present in the tissue sample. After neutralization, the extract is added to a light-yielding solution, and the luminescence is measured with a photomultiplier. The maximal height of the signal is measured by means of a digital voltmeter. The lumigen is bacterial luciferase with approximate additives and a supplement of malate and malate dehydrogenase.

The modified light-yielding solution provides for continuous formation of NADH (see Fig. 4.19), resulting in a durable level of light emission. The cycle involved was shown not to operate with $NADP^+$. The slow fading of the emission permits simplification of the measuring procedure.

### 2.1.8. Automated Fluorometric Analysis of Galactose in Blood (180)

In galactosemia, prevention of mental retardation depends on early recognition of the disorder and institution of dietary restriction of galactose. Mason et al. (180) described an automated fluorometric micromethod for galactose in whole blood spotted on filter paper. Galactose is oxidized by galactose oxidase to D-galactohexadialdose and $H_2O_2$ and is measured as the highly fluorescent condensation product of homovanillic acid (4-hydroxyl-3-methoxyphenylacetic acid; Fig. 4.32), formed when $H_2O_2$ is acted upon by horseradish peroxidase, as in Eq. (4.2):

$$\text{D-Galactose} + O_2 \xrightarrow{\text{galactose oxidase}} \text{D-galactohexodialdose} + H_2O_2$$

$$\text{Homovanillic acid} + H_2O_2 \xrightarrow{\text{peroxidase}} \text{fluorescent product} + H_2O$$

(4.2)

The procedure is tenfold more sensitive than colorimetric procedures for galactose and is not hampered by the nonspecific fluorescence from endogenous NADPH that is encountered in methods in which galactose dehydrogenase is used. The method has the requisite sensitivity and accuracy for quantification of galactosemia and galactosuria in milk-fed newborn infants and for genetic evaluation of the families of patients.

### 2.1.9. Evaluation of Fluorometrically Estimated Serum Bile Acid in Liver Disease (181)

Fasting serum bile acid (SBA) was measured by the enzymic fluorometric method coupled with the NAD-resazurin system in 23 controls, 35 asymptomatic carriers of hepatitis B surface antigen, including 4 e antigen carriers, and 91 patients with various liver diseases.

All hepatitis B surface and e antigen carriers showed SBA within the normal range. SBA was most significangly correlated with serum bilirubin ($\gamma = 0.74$) and was a more sensitive index for impaired liver function than bilirubin or alkaline phosphatase in 164 randomly chosen samples from the liver disease group. In serial determinations of SBA with reference to glutaminooxaloacetic transaminase (GOT), glutaminopyruvic transaminase (GPT), changing patterns of these two parameters were classified into the parallel type and the discrepant type. Of 40 cases with chronic liver disease, 32 belonged to the parallel type. SBA remained abnormal even after the normalization of transaminase in 12 out of 20 resolving episodes in cases of the parallel type, regardless of diagnosis. Since SBA changed according to the stage of the disease activity, serial and simultaneous estimation of SBA and of GOT-GPT was found to be helpful in the observation of liver diseases.

The principle of this method is as follows: Serum enzymes are inactivated by heating, bile acids are converted to 3-oxo bile acids with 3α-hydroxysteroid-oxidoreductase (3α-HSD) with concomitant reduction of NAD to NADH, and then the hydrogen of the generated

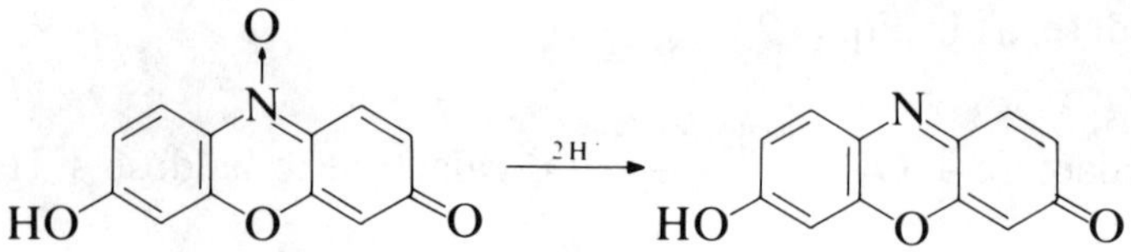


resazurin (nonfluorophor) resorfin (fluorophor)

**Figure 4.24.** Formation of resorfin from resazurin.

NADH is transferred by diaphorase to resazurin to yield the fluorophor, resorfin (7-hydroxy-3-isophenoxazone) (Fig. 4.24) (182, 183). Finally, the fluorescence of resorfin is measured; this is proportional to the amount of bile acids.

### *2.1.10. Fluorometric Determination of Plasma Unesterified Fatty Acid (184)*

Plasma unesterified fatty acid (UFA) has been determined by methods involving organic extraction into solvents and either titration (185) or formation of metal soaps and subsequent metal determination (186). The titration methods are accurate but time consuming and usually require large volumes of plasma. Metal soap methods require little plasma but suffer from many sources of error (187).

Curzon and Kantamaneni (184) describe a reliable and convenient method for plasma UFA analysis involving extraction by Dole's reagent (188) and determination by the decrease of fluorescence of a dilute solution of 7-hydroxy-4-methylcoumarin as a fluorescent indicator (Fig. 4.25). Results on rat and human plasma are in good agreement with those obtained by titration.

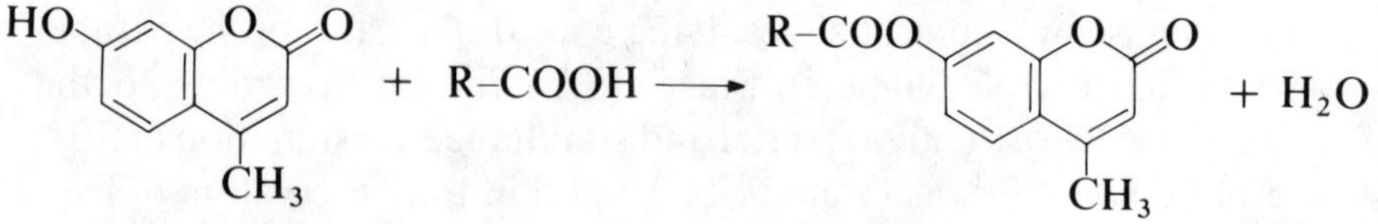


7-hydroxy-4-methylcoumarin (fluorophor) fatty acid 7-alkylcarboxy-4-methylcoumarin (nonfluorophor)

**Figure 4.25.** Nonfluorescent derivatization of fatty acid with 7-hydroxy-methylcoumarin as an extrinsic fluorophor.

## 2.1.11. New Fluorometric Method for Determination of Picomoles of Inorganic Phosphorus; Application to Renal Tubular Fluid (189)

Many of the methods available for the determination of inorganic phosphorus (Pi) have been based on the original technique of Bell and Doisy (190). These methods depend on the development of a specific color of the phosphomolybdate complex in the presence of reducing agents such as hydroquinone, 1,2,4-aminonaphtholsulfonic acid, ascorbic acid, and *p*-methylaminophenol sulfate.

Several micropuncture studies have utilized the colorimetric properties of phosphomolybdate to measure Pi in tubular fluid (191). However, despite the good sensitivity of these methods, a large volume of tubular fluid is required.

The method by Brunette et al. (189) is based on the reaction between phosphate, which is converted to hexadimolybdatophosphate, and thiamin, resulting in a highly fluorescent thiochrome (Figs. 4.26, 4.27) (192, 193). This technique allows the measurement of Pi in $2-4 \times 10^{-3}$ mm$^3$ of tubular fluid.

The reaction is stable and reproducible. Problems of interference are minimal. Microdeterminations of Pi in the proximal tubules and glomerular filtrates of the rat yielded results similar to those published in the literature.

## 2.1.12. Assay for Nanogram Quantities of DNA in Cellular Homogenates (194)

The fluorescent dye 4',6-diamidino-2-phenylindole (DAPI; Fig. 4.28) complexes with DNA to give a product with a fluoresence intensity about 20 times greater than that of the dye alone (195, 196).

Brunk et al. (194) measured accurately the DNA concentration of a crude cellular homogenate in the nanogram range using the fluorescence enhancement of DAPI or bisbenzimidazole (Hoechst H 33258).

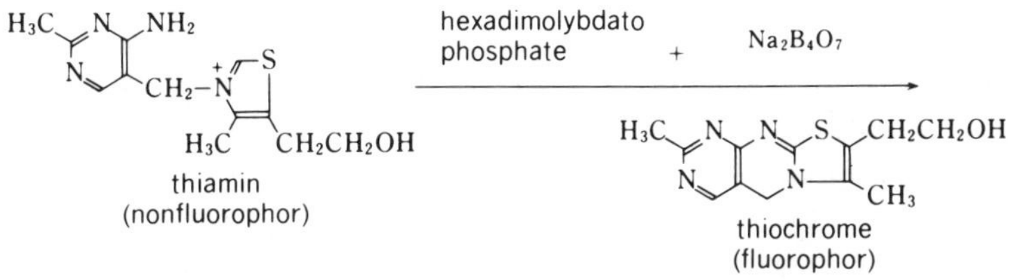

**Figure 4.26.** Oxdazing conversion of thiamin to fluorescent thiochrome (1).

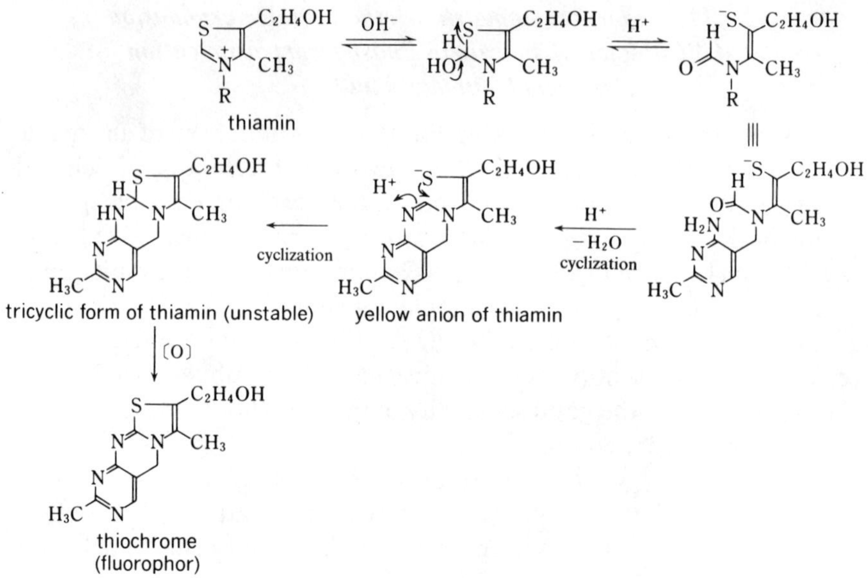

**Figure 4.27.** Oxdazing conversion of thiamin to fluorescent thiochrome (2).

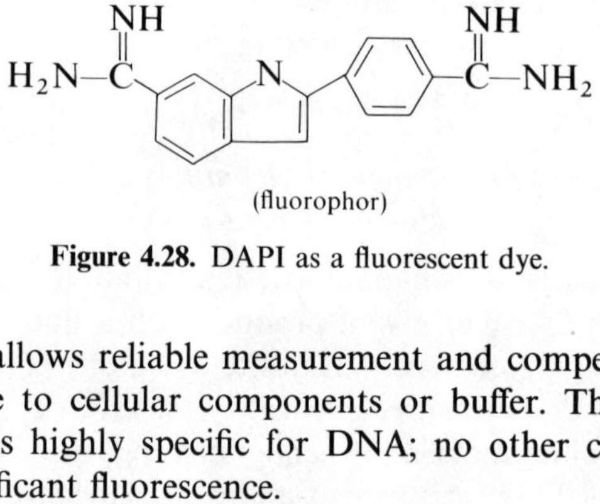

**Figure 4.28.** DAPI as a fluorescent dye.

This assay allows reliable measurement and compensates for any quenching due to cellular components or buffer. The fluorescence enhancement is highly specific for DNA; no other cell component produces significant fluorescence.

### 2.1.13. Intracellular pH Determination by Fluorescence Measurements (197)

To develop a method by which cells differing in internal pH ($pH_i$) could be separated using a fluorescence-activated cell sorter, Visser et al. (197) studied the possibility of using fluorescein as a pH indicator in mammalian cells.

This method is based on the observation that the fluorescence excitation spectrum of fluorescein is pH dependent. Fluorescence excitation spectra from individual rat-bone-marrow cells treated with

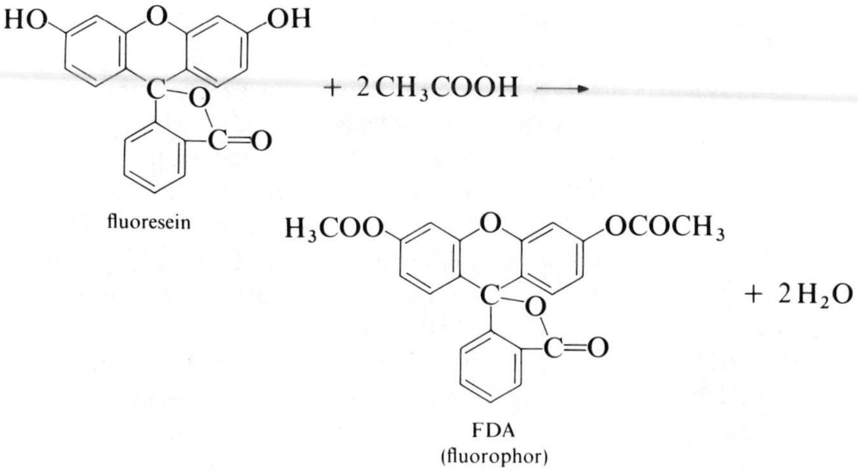

[Structural diagram: fluorescein + 2 CH₃COOH → FDA (fluorophor) + 2 H₂O]

**Figure 4.29.** Fluorescein diacetate as a fluorescent pH indicator.

fluorescein diacetate (FDA; Fig. 4.29) were compared with those of fluorescein solutions of known pH values.

### 2.1.14. Detection of Diamagnetic Cation in Tissue Using the Fluorescent Probe Chlorotetracycline (198)

Minute quantities of metallic cations are essential for most biological systems to function properly, and some cations (e.g., $Ca^{2+}$ and $Mg^{2+}$) tend to accumulate in high concentration in small, selected areas, especially associated with biological membranes (199). X-ray microanalysis is a useful technique in analyzing these areas. However, with this technique alone, it is time consuming and difficult to locate the required areas. Macinnes et al. (198) have attempted to solve the problem by using chlorotetracycline (Fig. 4.30) as a fluorescent probe. Chlorotetracycline has been used as a membrane fluorescent agent since it forms chelate complexes with divalent diamagnetic cations,

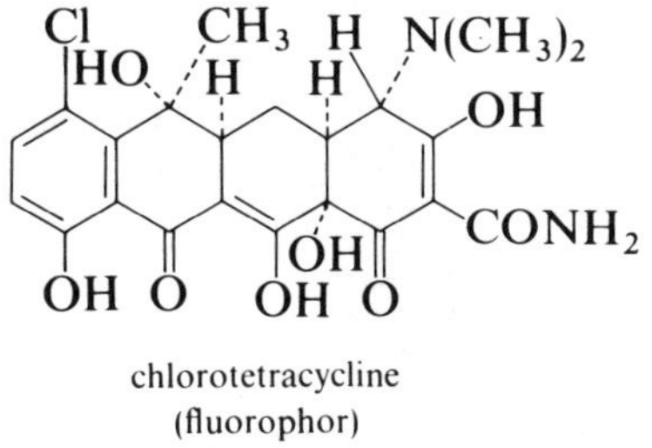

[Structural diagram of chlorotetracycline (fluorophor)]

**Figure 4.30.** Chlorotetracycline as a fluorescent probe.

giving enhanced emission spectra when excited at specific wavelengths.

The present work has used chlorotetracycline as a probe to stain human brain postmortem from a normal subject and from a case of presenile dementia of the Alzheimer type. When stained tissue is examined in the cathodoluminescent mode (200) in a scanning electron microscope, a low-intensity luminescent background is observed on which several very intense spots are visible. This luminescence is emitted from very small areas in the tissue (diameter 0.5–2 $\mu$m) and is so intense that it enables a single 0.5-$\mu$m-diameter area to be localized in the cathodoluminescent mode using a magnification of only × 30.

Once located, these areas were analyzed using energy-dispersive X-ray microanalysis and were found to be associated with sulfur in combination with heavy metals (e.g., zinc and cadmium). This suggests discrete areas of a metal-protein complex such as occur, for example, in well-known enzyme structures like those of carboxypeptidase and carbonic anhydrase.

### 2.1.15. Fluorometric Microassay of DNA Using a Modified Thiobarbituric Acid Assay (201)

A method for the rapid, specific measurement of minute amounts of DNA is presented by Nordling and Aho (201). The method is sensitive to about 20 ng of DNA, which is equivalent to approximately 3000 mammalian cells, and employs enzymatic liberation of deoxyribose with DNase and phosphodiesterase. The liberated deoxyribose is reacted with thiobarbituric acid to yield a fluorescing compound

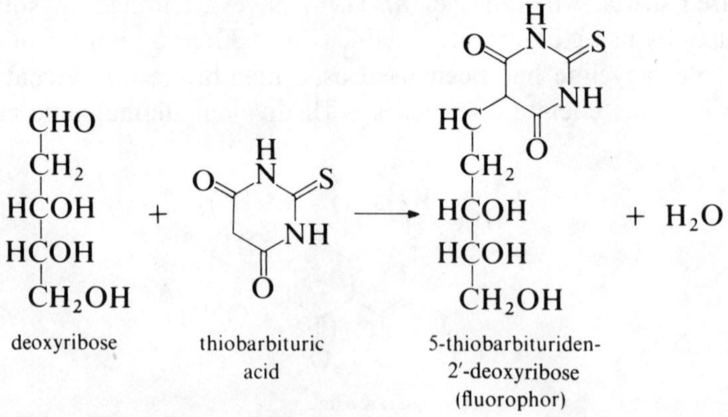

**Figure 4.31.** Fluorescent derivatization of deoxyribose with thiobarbituric acid as a fluorogenic reagent.

(Fig. 4.31) with an excitation maximum at 532 nm and an emission maximum at 549 nm, measured with a fluorometer. The entire procedure is accomplished within a few hours. The result is linear up to about 1 μg DNA or $3 \times 10^5$ lymphocytes.

### 2.1.16. Fluorometric Oxidase Assays: Pitfalls Caused by the Action of Ultraviolet Light on Lipids (202)

A number of fluorometric assays of hydrogen peroxide-producing oxidases are based on the formation of highly fluorescent products from homovanillic acid (4-hydroxy-3-methoxylphenylacetic acid, HVA) or related compounds by horseradish peroxidase. Homovanillic acid is oxidized by hydrogen peroxide and horseradish peroxidase (HRP) to form a highly fluorescent compound, 5,5'-bis(4-hydroxy-3-methoxy-1-phenylacetic acid) (Fig. 4.32) (203, 204). This reaction forms the basis for a sensitive assay for hydrogen peroxide first proposed by Guilbault et al. (204).

Hirsch and Parks (202) observed that under continuous UV illumination at the wavelengths used for excitation in these methods, brain or muscle homogenates produce fluorescence increases in the absence of any exogenous enzyme substrate; when UV light is excluded, however, such increases are negligible. Arachidonic and linolenic acids also produce this effect. For this reason, measurements of $H_2O_2$ based on this principle are valid only if this nonspecific effect has been excluded, and should preferably be carried out as endpoint rather than continuous assays. It is believed that the effect of UV light on the reaction is due to formation of $H_2O_2$ and/or oxygen free radicals, and polyunsaturated lipids appear to be involved as intermediates. Thus, the homovanillic acid-horseradish peroxidase system may prove useful in investigations of the effect of UV on the production of oxygen free radicals and lipid peroxidation.

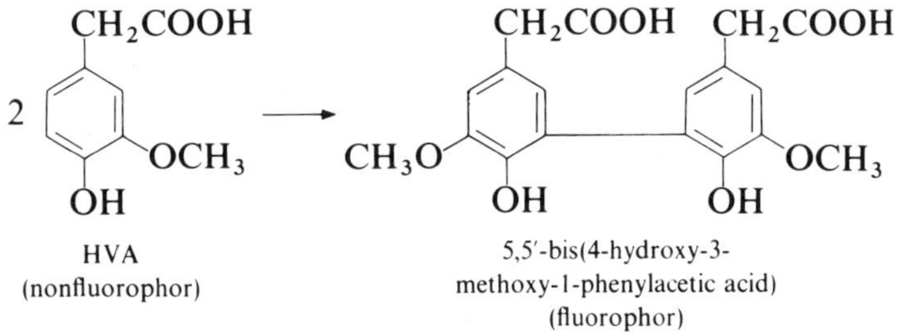

**Figure 4.32.** Formation of fluorophor by dimerization of HVA.

### 2.1.17. Fluorometric Analysis: A Study on Fluorescent Indicators for Measuring Near-Neutral ("Physiological") pH Values (205)

The desired properties of fluorescence pH indicators are briefly summarized by Wolfbeis et al. (205). Fluorescence maxima and $pK_a$ values of 28 potential indicators are presented in detail, while other data (fluorescence quantum yields, solubility, and stability) are given qualitatively.

Among the indicators investigated, some coumarins and HPTS (1-hydroxy-pyrene-3,6,8-trisulfonate; Fig. 4.33) are considered to be the most suitable for both acid-base titrations and precise pH measurements. The HPTS absorption, excitation, and fluorescence spectra, decay times, $pK_a$ values, and fluorescence properties as a function of pH and excitation wavelength are given. As a result of its stability, water solubility, long-wave excitation maximum, and large Stokes shift, HPTS appears to be the pH indicator of choice.

### 2.1.18. Prolidase Deficiency: Characteristics of Human Skin Fibroblast Prolidase Using Colorimetric and Fluorometric Assays (206)

Prolidase deficiency is an autosomal recessive disease associated with chronic ulcerative dermatitis, mental retardation, and iminodipeptiduria (207). A single form of prolidase was found in human erythrocytes (208), and an almost total deficiency of the enzyme against the substrate glycylproline has been reported (209) in the disease. In a preliminary report on cultured human skin fibroblast prolidase (210), Priestman and Butterworth (206) showed that the disease enzyme activity against other substrates was not as reduced. Characteristics of the normal and abnormal enzyme of skin fibroblasts are described that make it possible to diagnose the disease using other substrates in addition to glycylproline.

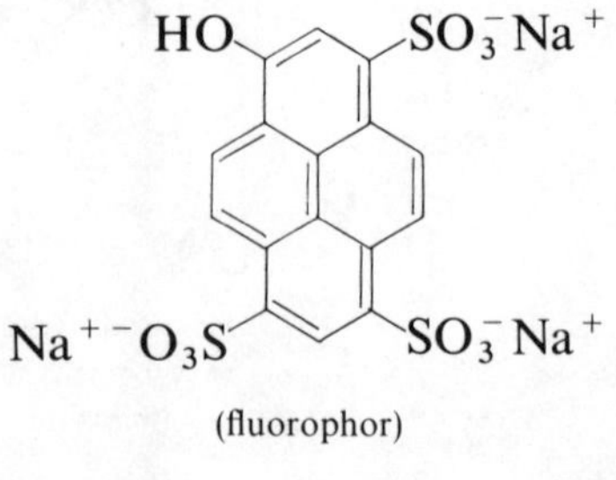

(fluorophor)

**Figure 4.33.** HPTS as a fluorescent pH indicator.

The principle of the fluorometric assay of prolidase activity is as follows: Prolidase is reacted with its substrate, prolylamino acid, to liberate animo acid, which is reacted with L-amino acid oxidase. Hydrogen peroxide formed in this oxidase reaction is measured fluorometrically by converting homovanilic acid to a highly fluorescent compound in the presence of peroxidase (see Fig. 4.32) (211).

## 2.2. Immunology (212, 213)

Immunological methods have rapidly become standard practice in clinical chemistry because of their specificity, potential sensitivity, practicability, and wide applicability. In particular, they are used for measuring biologically active compounds that are present in very low concentrations, such as proteins (enzymes, receptors, antibodies), hormones (steroids, thyroid hormones, peptide hormones), drugs, and microorganisms.

Radioimmunoassays are the most widely used immunological assays. Although extremely sensitive and quite precise, these assays have several drawbacks. Because radiation may be a health hazard, special attention to the handling of reagents, the training of staff, and the storage of waste is required. The useful lifetime of a kit is limited by the half-life of the isotope. Counting radioactivity requires special, expensive instrumentation and is time consuming. For these reasons alternative analytical methods have been developed. Among the alternative labels developed to substitute for radiotracers are enzymes (214), spin labels (215), luminescent compounds (216), metals (217), and fluorescent probes (218).

Fluoroimmunoassays (219–226) based on labeling immunoreactants with fluorescent probes, are rapidly gaining wider applications. Although fluoroimmunoassay as a quantitative method is quite new, labeling antibodies with fluorescent probes has been used as a qualitative staining technique since 1941 (227). The high sensitivity of the fluorescence measurement, combined with the sensitivity of the probe to changes in its enviroment, offers possibilities for developing heterogeneous assays and, above all, simple and rapid homogeneous assays—assays where the concentration of analyte can be monitored directly in the reaction mixture.

The problem with fluoroimmunoassay methods has been their inferior sensitivity, caused, to a great extent, by the high background of the fluorometric measurement. The development of solid-phase separation systems, new fluorescent probes, and new instrumentations (e.g., time-resolving measurement) has lowered the background so that sensitivity is now comparable to that of radioimmunoassays.

### 2.2.1. Types of Fluoroimmunoassay

Like other immunoassays, fluoroimmunoassays (FIA) are based on the interaction of an antigen with its specific antibody; the assay principles, reactions, separation of free fraction from bound fraction, and so on are the same. Fluorescence, however, offers some additional advantages for developing homogeneous assays, in which the separation step is not needed.

FIAs can be categorized in several ways, for example, those with labeled antigens versus those with labeled antibodies, or those featuring competition between antigens (tracer and free) for a limited amount of antibodies (like radioimmunoassays) versus noncompetitive assays (excess reagent methods, e.g., "sandwich" assays). In this discussion we have classified the assays as heterogeneous or homogeneous, depending on whether the separation step is or is not needed.

**2.2.1.1. Heterogeneous Fluoroimmunoassays.** In heterogeneous assays free labeled antigens must be separated from antibody-bound antigens, or unbound labeled antibodies from an antigen-bound fraction, before measurement. As in radioimmunoassays, this can be accomplished by precipitation or by use of solid-phase-bound antigens or antibodies. These assays are called *separation fluorommunoassays* (Sep-FIAs) (228–236). More often, however, solid-phase-based sandwich assays with labeled antibodies are used. In analogy with immunoradiometric assays (IRMA), these latter assays are called *immunofluorometric assays* (IFMA) (237). The development of solid-phase types and of materials with low fluorescence backgrounds enhances sensitivity and avoids the problems of background fluorescence from serum.

**2.2.1.2. Homogeneous Fluoroimmunoassays.** In homogeneous fluoroimmunoassays the antibody-bound antigen need not be separated from the free antigen before the fluorescence measurement. The analyte concentration in a sample can be monitored directly from the reaction mixture through its effects on the fluorescence properties of the labeled antigen or antibody. Thus, the assays are very rapid and simple (one incubation and no washings), and their development has been among the main objectives in FIA research. However, the sensitivity of homogeneous assays is often seriously limited by interferences from samples (serum) and by the low degree of fluorescence change (quenching, enhancement, polarization, energy transfer) in an immunoreaction. Most of the applications of homogeneous FIA methods are designed for analytes present in substantial concentrations, for example, drugs.

Homogeneous fluoroimmunoassays are further classified as follows: fluorescence polarization immunoassays (FPIA) (238–243), fluorescence enhancement immunoassays (244–247), fluorescence quenching immunoassays (248–249), fluorescence excitation transfer immunoassays (FETIA) (250–256), release fluoroimmunoassays (257–259), and homogeneous fluorescence immunoassays based on special instrumentation and so on (260–263).

The properties and structures of some probes used in fluoroimmunoassays are shown in Table 4.10 and Fig. 4.34, respectively.

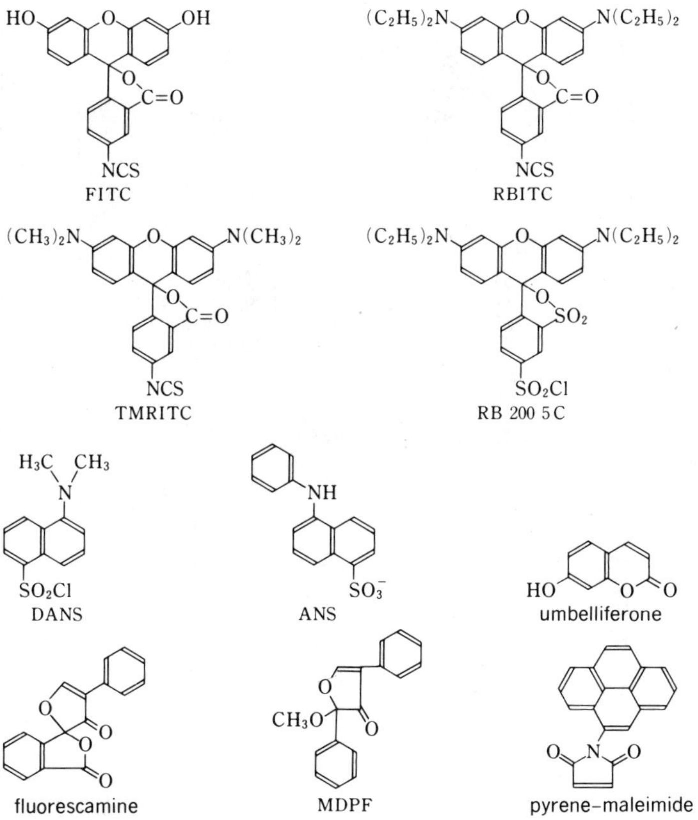

**Figure 4.34.** Structures of some fluorescent probes used in fluoroimmunoassays (see abbreviations in Table 4.10). [Reprinted with permission from I. Hemmilä, *Clin. Chem.*, **31**, 359 (1985).]

Table 4.10. Properties of Some Fluorescent Probes Used in Fluoroimmunoassays

| Probe | $\lambda_{abs/ex}$ (nm) | $\varepsilon$ | $\lambda_{em}$ (nm) | Quantum Yield (%) |
|---|---|---|---|---|
| Fluorescein (FITC, DTAF) | 492 | $7 \times 10^4$ | 520 | 0.85 |
| Rhodamines | | | | |
|   RBITC | 550 | $1.2 \times 10^4$ | 585 | 0.70 |
|   TMRITC | 550 | $5.0 \times 10^4$ | 580 | |
|   RB 200 SC | 530, 565 | | 595 | |
| Umbelliferones | 380 | $2.0 \times 10^4$ | 450 | |
| DANS | 340 | $3.4 \times 10^3$ | 480–520 | 0.30 |
| ANS | 385 | | 471 | 0.80 |
| Fluorescamine | 394 | $6.3 \times 10^3$ | 475 | 0.10 |
| MDFP | 390 | $6.4 \times 10^3$ | 480 | 0.10 |
| Pyrene deriv.[a] | 340 | | 375, 392 | |
| Lucifer Yellow VS | 430 | | 540 | |
| Porphyrins | 400–410 | | 619–633 | |
| Chlorophylls | 430–453 | | 648–669 | |
| Phycobiliprotein | 550–620 | $7.0 \times 10^5$ | 580–660 | 0.50–0.98 |
| Eu-($\beta$-NTA)$_3$ | 340 | $2.4 \times 10^6$ | 590, 613 | |
| Tb-EDTA-sulfosalicyclic acid | 300 | $3 \times 10^4$ | 490, 545 | |
| Nd-benzoyltrifluoroacetone | 800 | | 900, 1060, 1350 | |

[a] $N$-($\beta$-Pyrene)maleimide.

Abbreviations: NTA, naphthoyltrifluoroacetone; DTAF, dichlorotriazinylaminofluorescein; MDFP, 2-methoxy-2,4-diphenyl-3(2H)-furanone; DANS, dansyl chloride; ANS, anilinonaphthalenesulfonic acid; RBITC, rhodamine B isothiocyanate; TMRITC, tetramethylrhodamine isothiocyanate; RB 200 SC, lissamine rhodamine B sulfonyl chloride.

Source: Reprinted with permission from I. Hemmilä, Clin. Chem., **31**, 359 (1985).

### 2.2.2. Laser Fluorescence Immunoassay of Insulin (264)

Recently, the use of laser excitation in fluorescence analysis has opened new possibilities for ultrasensitive detection in the liquid phase (265–268). Also, the combination of laser fluorometry with fluorescence immunoassay has been attempted in order to achieve an attractive alternative to radioimmunoassay.

Immunoassay is based on the competitive reaction of unlabeled antigen (Ag) with labeled antigen (Ag*) for binding sites on an antibody (Ab), directed against the antigen, as shown in Eq. (4.3):

$$Ag^* + Ag \rightleftharpoons Ag^*Ab$$
$$+$$
$$Ag \qquad (4.3)$$
$$\downarrow\uparrow$$
$$AgAb$$

In most cases, the fluorescences from Ag* and Ag*Ab are not sufficiently different in polarization and/or wavelength, and the background fluorescence of the solvent overwhelms the signal of interest. Under such conditions, it is necessary to resort to a separation procedure for Ag* and Ag*Ab, for example, chromatography, electro-

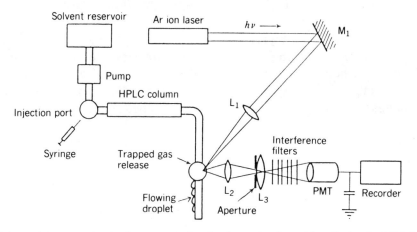

**Figure 4.35.** Experimental apparatus for the determination of insulin by the HPLC separation on the basis of molecular size with a gel filtration column and the excitation with a CW argon ion laser. [Reprinted with permission from S. D. Lidofsky, T. Imasaka, and R. N. Zare, *Anal. Chem.*, **51**, 1602. Copyright 1979 American Chemical Society.

phoresis, or physical adsoprtion onto a solid phase (heterogeneous immunoassay).

Lidofsky et al. (264) report a method for the immunoassay of the protein hormone insulin in aqueous buffers. Fluorescein isothiocyanate (FITC; see Fig. 4.34) serves as the antigen label, and a CW argon-ion laser is used as a fluorescent excitation source (Fig. 4.35). Insulin bound in complex is separated from free insulin on the basis of molecular size by using a gel filtration column in a HPLC system. FITC is a versatile compound for labeling proteins as it binds to free amino groups (269). It has a large absorption coefficient ($\varepsilon_{490} \approx 7.66 \times 10^4$) (270), a high fluorescence quantum yield ($\phi \approx 0.8$), and long-term stability (269). Its excitation maximum occurs at 490 nm, which nearly coincides with the strong 488 nm line of the argon-ion laser. The emission maximum of FITC occurs at 520 nm (271). Its visible spectrum does not appreciably change on binding to protein, but the fluorescent quantum yield is reduced (271).

### 2.2.3. Luminescence Immunoassay of Human Serum Albumin with Hemin as Labeling Catalyst (272)

An unique, nonradioactive solid-phase immunoassay, which uses a small molecular catalyst to label an analyte, is described for the determination of organic substances in solution by Ikariyama et al. (272). The catalyst, bound covalently to the analyte, can be determined with high sensitivity by virtue of its catalytic activity in a manner similar to that in enzyme immunoassay. This concept has been demonstrated by using hemin, which catalyzes the luminescence reaction of luminol and hydrogen peroxide, as the label and human serum albumin (HSA) as the analyte (Fig. 4.36). When hemin is conjugated to HSA, catalytic activity is retained in free solution and when bound to HSA antibody. A constant amount of hemin-labeled HSA is added to the test solution in the presence of an antibody-bound plate. Nonlabeled and hemin-labeled HSA compete for binding to the plate-bound antibody. After thorough washings, the plate is contacted with a solution containing luminol and hydrogen peroxide and assayed for hemin by measuring luminescence intensity [see Eq. (4.4)]:

$$\text{Luminol} + H_2O_2 \xrightarrow{\text{hemin}} \text{aminophthalate} + N_2 + H_2O + h\nu$$

(4.4)

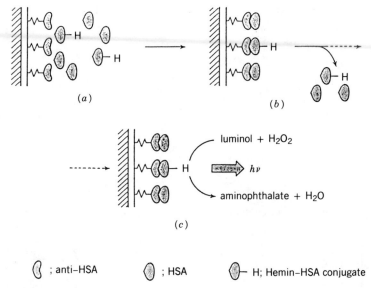

**Figure 4.36.** Schematic representation of hemin catalyst immunoassay of human serum albumin (HSA): (a) competitive immunoreaction of nonlabeled and hemin-labeled HSA with plate-bound antibody; (b) removal of unreacted nonlabeled and hemin-labeled HSA; (c) chemiluminescence reaction of luminol and $H_2O_2$ catalyzed by bound hemin-labeled HSA. [Reprinted with permission from Y. Ikariyama, S. Suzuki, and M. Aizawa, *Anal. Chem.*, **54**, 1126. Copyright 1982 American Chemical Society.]

### 2.2.4. Chemiluminescence-Labeled Antibodies and Their Applications in Immunoassays (273)

Chemiluminescent molecules fulfil the criteria for use as labels in immunoassay. They are readily available, can be quantified using relatively simple equipment, and can be coupled to protein to produce stable derivatives, which retain both chemiluminescent and immunological activity. Their unique advantage over other nonisotopic labels is their high sensitivity of detection, acridinium esters being readily quantified in amounts as little as $10^{-17}$ mol or less.

Woodhead et al. (273) used acridinium-ester-labeled rabbit immunoglobulin G (IgG) in a system analogous to that in conventional radioimmunoassay. Label and standards were reacted for 18 h with a solid-phase antibody consisting of a sheep antirabbit IgG preparation linked covalently to reprecipitated cellulose. Separation way by centrifugation, and the bound label was quantified by chemiluminescent reaction (Fig. 4.37).

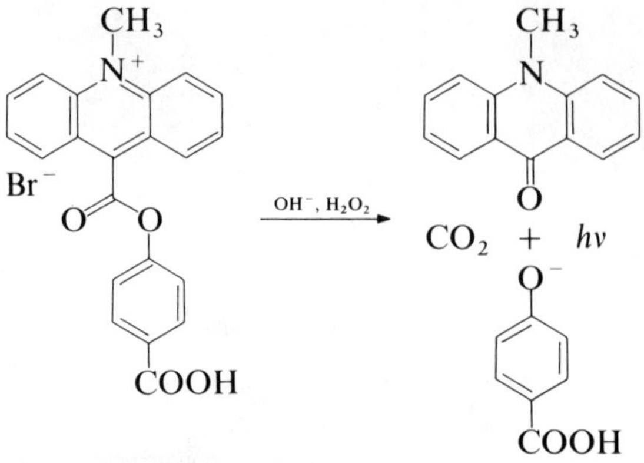

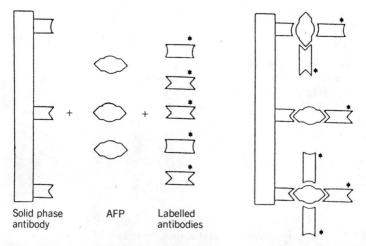

**Figure 4.37.** Chemiluminescent reaction of an acridium ester.

**Figure 4.38.** Two site luminometric assay protocol. [Reprinted from J. S. Woodhead, et al., in *Luminescent Assays: Perspective in Endocrinology and Clinical Chemistry*, in Serio and M. Pazzagli (Eds.), Raven Press, New York, p. 147 (1982).]

A two-site assay of human α-fetoprotein (AFP) was set up using the protocol outlined in Fig. 4.38. Standards were reacted simultaneously with cellulose-linked sheep anti-AFP antibody and labeled antibody for 1 h at room temperature. After centrifugation the bound chemiluminescence was measured.

### 2.2.5. *Acridinium Esters as High-Specific-Activity Labels in Immunoassay (274)*

A chemiluminescent acridinium ester (see Fig. 4.37) has been synthesized that reacts spontaneously with proteins to yield stable, immunoreactive derivatives of high specific activity. The compound has been used to prepare chemiluminescent monoclonal antibodies to human $\alpha_1$-fetoprotein having average incorporation ratios as great as 2.8 mol of label per mole of antibody, which corresponds to a detection limit of approximately $8 \times 10^{-19}$ mol. These antibodies have been used in the preliminary development of a two-site immunochemiluminometric assay for human $\alpha_1$-fetoprotein, which requires only a 30-min incubation and a quantification time of 5 s per sample.

### 2.2.6. Two-Site Immunochemiluminometric Assay for Human $\alpha_1$-Fetoprotein (275)

In this two-site immunochemiluminometric assay for human $\alpha_1$-fetoprotein, acridinium-ester-labeled monoclonal antibodies are used that have an average incorporation ratio of 0.3 mol of acridinium ester per mole of antibody. The solid-phase antibody consists of sheep anti-$\alpha_1$-fetroprotein IgG covalently coupled to a diazonium derivative of reprecipitated aminoaryl cellulose. The assay involves a 1-h incubation after the simultaneous incubation of labeled and solid-phase antibodies.

Experiments with monoclonal antibodies labeled to an average incorporation ratio of 2.8 mol of acridinium ester per mole of antibody suggest that assay sensitivity increases with increased specific activity. $\alpha_1$-Fetoprotein concentrations in the sera of pregnant women at 14–20 weeks of gestation, as measured by the present assay, agreed with the results of a conventional radioimmunoassay.

### 2.2.7. Homogeneous Immunoassay Based on Chemiluminescence Energy Transfer (276)

The chemiluminescent compound aminobutylethyl isoluminol (ABEI) and its isothiocyanate derivative have been coupled to a range of haptens (progesterone, cyclic AMP, cyclic GMP) and protein antigens (IgG, C9) (Fig. 4.39). All the derivatives were chemiluminescent, immunologically active, and stable for more than 6 months. When the ABEI-labeled antigens bind to their respective fluorescein-labeled antibodies, there is a shift in the ratio of chemiluminescence at 460 nm (blue) to that at 525 nm (green). This nonradioactive energy transfer was used to establish homogeneous immunoassays, which require no separation step. These assays were at least as sensitive as the conventional radioimmunoassays and could accurately measure substances in serum (i.e., for IgG, results correlated by $\gamma = 0.96$ with those from $^{125}$I assay) and tissue extracts (cyclic AMP, $\gamma = 0.91$ relative to

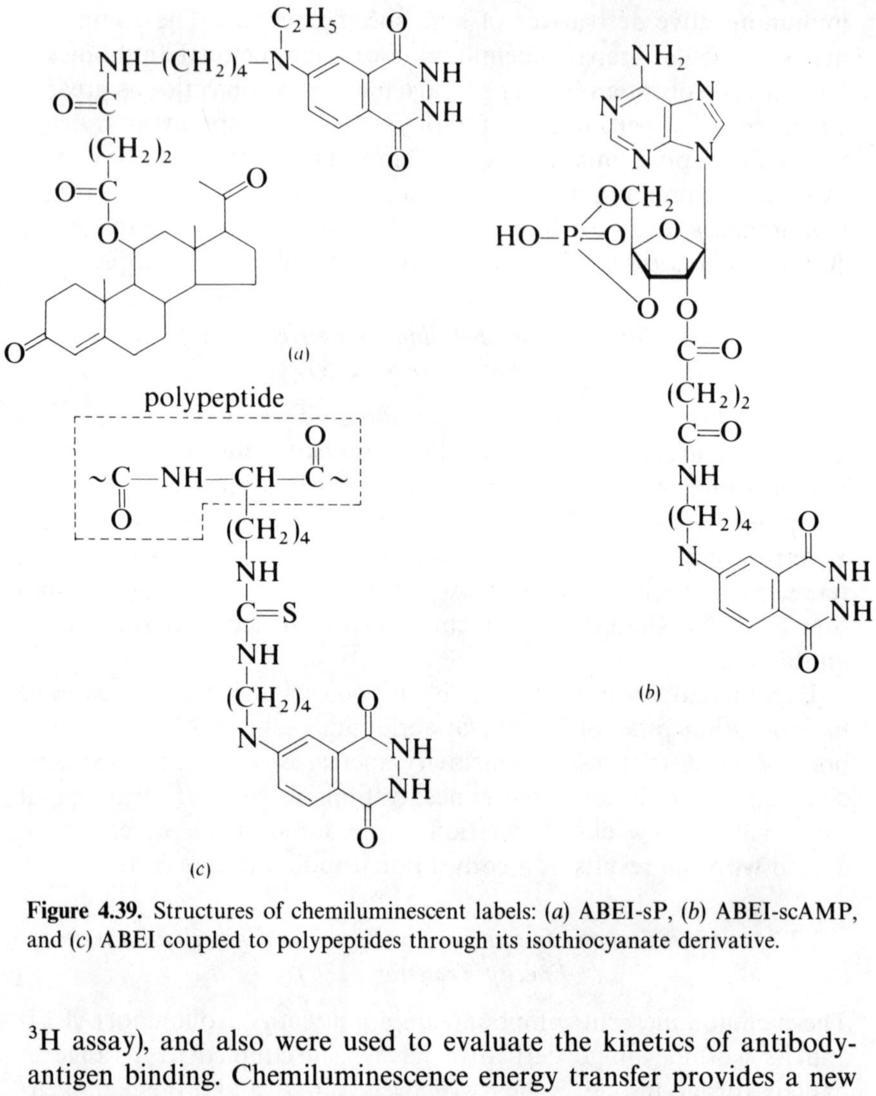

**Figure 4.39.** Structures of chemiluminescent labels: (a) ABEI-sP, (b) ABEI-scAMP, and (c) ABEI coupled to polypeptides through its isothiocyanate derivative.

$^3$H assay), and also were used to evaluate the kinetics of antibody-antigen binding. Chemiluminescence energy transfer provides a new method for quantifying ligand-ligand interactions in the $10^{-15}$ to $10^{-18}$ mol range without first separating bound and free ligand. This development provides a unique opportunity to investigate chemical events in single cells and intact cells.

### 2.2.8. Enhanced Luminescence Procedure for Sensitive Determination of Peroxidase-Labeled Conjugates in Immunoassay (277)

The present luminescence assay for horseradish peroxidase (HRP) has limitations. Therefore, Whitehead et al. (277) have reported a novel procedure in which the HRP-catalyzed luminescence of a cyclic hydrazide, such as luminol, is multiplied severalfold by the addition of a synthetic component of the firefly bioluminescent system, D-luciferin (4,5-dihydro-2-(6-hydroxy-2-benzothioazolyl)-4-thioazole-carboxylic acid). The specific enhancement of HRP-catalyzed light emission from cyclic hydrazides should extend the sensitivities of luminescently monitored assays, which have already been shown to be as sensitive as those using radioactive labels. This procedure has been applied to the immunoassay of serum α-fetoprotein, thyroxine, digoxin, hepatitis B surface antigen, immunoglobulin E, and rubella virus antibody.

### 2.2.9. Chemiluminescent Tags in Immunoassays (278)

Although radioimmunoassay procedures have a number of advantages, they do pose problems, especially with regard to the disposal of radioactive waste. Alternative methods, such as chemiluminescence, and enzyme, fluorescence, or spin immunoassays, have been tested by Klingler et al. (278) with the aim of replacing radioactive labels without loss of sensitivity, precision, or accuracy.

Chemiluminescence immunoassays using luminogenic-compound-labeled antigens have been described in different publications; the most important methods are listed in Table 4.11. The chemilumines-

**Table 4.11. Chemiluminescence Immunoassays Using Luminogenic Compound-Labeled Antigens**

| Antigen | Luminogenic Compound | Reference |
|---|---|---|
| Progesterone | ABEI | 280 |
|  | ABEI, AHEI | 281, 282 |
| Cortisol | ABEI | 283 |
|  | ABEI, ABI | 284, 285 |
|  | APEI | 286 |
| Thyroxine | ABEI | 287 |
| Pregnandiol-3α glucuronide | AHEI | 288 |
| Estriol-16-glucuronide | ABEI | 289 |
| Testosterone | ABEI | 290 |
| Unconjugated estriol | ABEI | 291 |
| Total estriol | ABEI | 292 |

Abbreviations: ABEI, aminobutylethylisoluminol; AHEI, aminohexylethylisoluminol; ABI, aminobutylisoluminol; APEI, aminopentylethylisoluminol.

cence label of choice in immunoassays is at present 6 [*N*-(4-aminobutyl)-*N*-ethyl]amino-2,3-dihydrophthalazine-1,4-dione (aminobutylethylisoluminol, ABEI). Its synthesis has been described by Schroeder et al. (279); Fig. 4.40.

A new alternative to radioimmunoassay procedures is the solid-phase antigen luminescence technique (SPALT) that Klingler et al. (278) are currently developing. The general principle is shown in Fig. 4.41. The first incubation of antigen with first antibody is performed as in a radioimmunoassay, with "cold" preincubation. An excess of solid-phase antigen is then added to react with the free antibody. After washing and centrifuging, followed by removal of the supernatant, the solid phase is incubated with luminescence-labeled second antibody. The results of preliminary experiments using bioluminescent markers have recently been published, although replacement of the bioluminescent enzyme label (pyruvate kinase) with azoluminol, together with a suitable oxidation system, allows chemi-

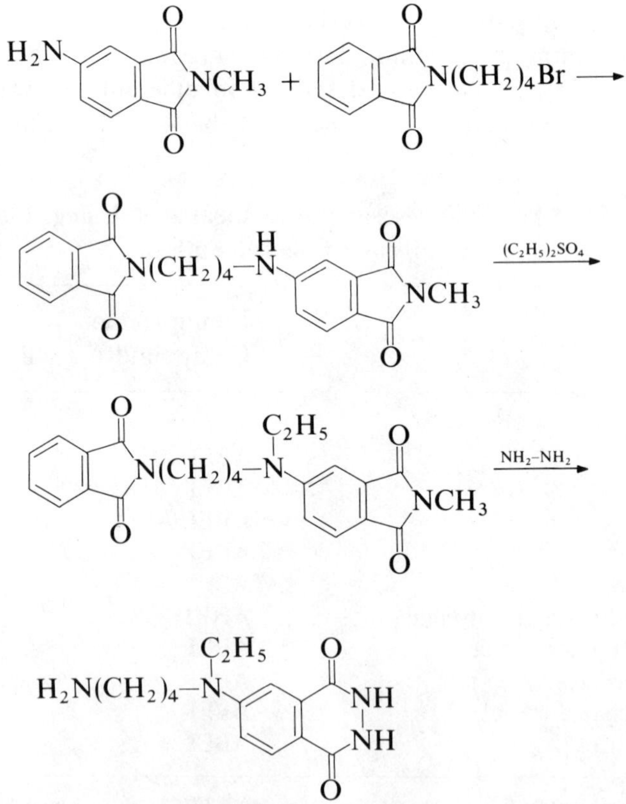

**Figure 4.40.** Synthetic pathway of ABEI.

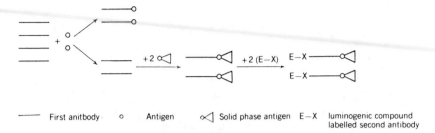

Figure 4.41. General principle of the solid phase antigen luminescence technique (SPALT). [Reprinted from W. Klinger, C. J. Strasburger, and W. G. Wood, *Trends in Analytical Chem.*, **2**, 132 (1983).]

luminescent solid-phase antigen luminescence techniques to be effected.

### 2.2.10. Direct Solid-Phase Fluoroenzymeimmunoassay of 5β-Pregnane-3α,20α-diol-3α-glucuronide in Urine (293)

A competitive solid-phase fluoroenzymeimmunoassay has been developed by Shah et al. (293) for rapidly measuring 5β-pregnane-3α,20α-diol-3α-glucuronide (Pd-3G) directly in diluted specimens of prebreakfast urine. The assay involves use of an antiserum to Pd-3G and enzyme-labeled antigen prepared by chemically linking glucose-6-phosphate dehydrogenase (EC 1.1.1.49) to Pd-3G. Antibody-bound antigen and free antigen are separated by use of a solid-phase double antibody: sheep antirabbit gammaglobulin coupled to cellulose particles. The solid phase, isolated by centrifugation, is washed free of labeled antigen and endogenous enzyme interferences. Enzyme activity in the bound fraction is then measured fluorometrically with glucose-6-phosphate as the substrate and generated NADPH (Fig. 4.42) as a fluorogenic indicator of enzyme activity. The assay is sufficiently sensitive (30–35 pg per assay tube), specific, and reliable for routine use, and results correlate well ($\gamma = 0.98$) with those from an established specific radioimmunoassay. Because this assay is suited to routine use, it may be applied to detecting ovulation, assaying the function of the corpus luteum, and monitoring early pregnancy.

### 2.2.11. On-Line Computer Analysis of Chemiluminescent Reactions, with Application to a Luminescent Immunoassay for Free Cortisol in Urine (294)

Tommasi et al. (294) interfaced a microcomputer on-line with a luminometer to acquire the light signal of chemiluminescent reactions

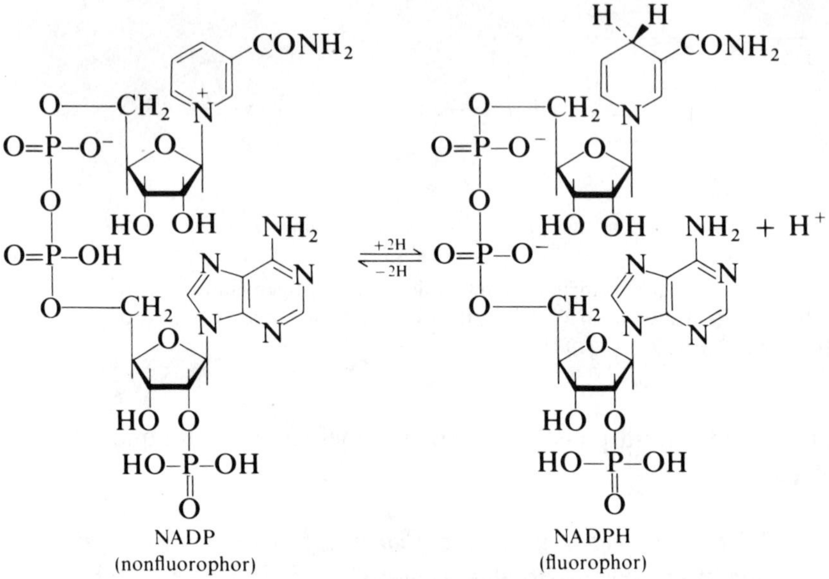

**Figure 4.42.** Oxidation–reduction reaction between NADP and NADPH.

from a photomultiplier and then compute significant parameters of light emission and kinetic "shape" indices. Using this system to study interferences from biological samples on the measurement of chemiluminescent reactions, the present authors observed that such effects are usually associated with modifications of the shape of the light-emission kinetics. These results suggest that simultaneous evaluation of the shape of a chemiluminescent reaction and a measurement of light emission can be combined to assess a luminescent immunoassay as an internal control of the interferences in measurements of the chemiluminescent tracer. As an example of this approach, the present authors developed and validated a luminescent immunoassay for free cortisol in diluted urine. Dextran-coated charcoal is used for bound-free separation.

### 2.2.12. Solid-Phase Chemiluminescence Immunoassay for Progesterone in Unextracted Serum (295)

Boever et al. (295) describe a simple, solid-phase chemiluminescence immunoassay for progesterone in 10 mm$^3$ of unextracted serum ("direct" assay). Danazol at pH 8.0 is included (100 ng per tube) to displace progesterone from binding proteins in serum. A progesterone-11α-hemisuccinyl-ABEI (see Table 4.11 and Fig. 4.40) conjugate serves as the chemiluminescent ligand marker, and homologous antiprogesterone IgG covalently coupled to "immunobeads" is the immunoadsorbant. After the binding reaction, bound ligand and free ligand are separated by centrifugation and the chemiluminescence yield of the bound label is determined. The sensitivity, specificity, precision, and accuracy of the method are similar to those of a conventional radioimmunoassay for progesterone, in which a radioligand of tritiated progesterone and serum extraction are used. Progesterone values obtained by this procedure agreed well ($\gamma = 0.987$) with those obtained by radioimmunoassay. The present authors conclude that chemiluminescence immunoassay for progesterone in unextracted serum is analytically valid and offers a convenient alternative to radioimmunoassay.

### 2.3. Medical Diagnoses

#### 2.3.1. Fluorometric Determination of δ-Aminolevulinate Dehydratase Activity in Human Erythrocytes as an Index of Lead Exposure (296)

Chakrabarti et al. (296) describe a fluorometric method for determining δ-aminolevulinate dehydratase (ALA-D; EC 4.2.1.24) activity in human erythrocytes and compare it with the existing colorimetric methods (297–301). Incubation conditions are identical. In the proposed method, the porphobilinogen (PBG) formed during incubation is converted to a stable and highly fluorescent uroporphyrin (Fig. 4.43) by heating for 20 min at 93°C in acidic medium in the presence of

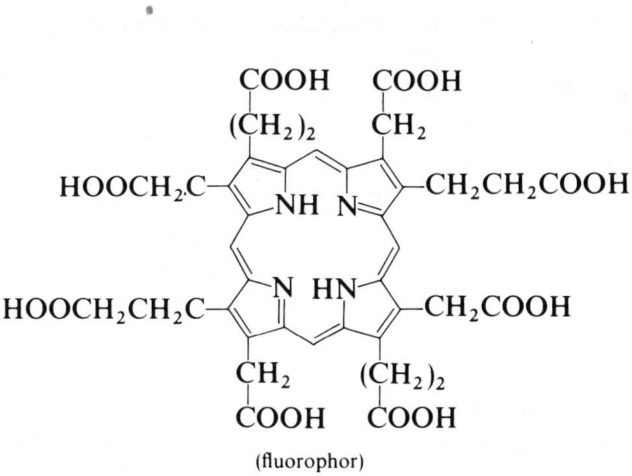

**Figure 4.43.** Structure of uroporphyrin.

air. The correlation between results by the two methods was good ($\gamma = 0.89$).

The present method is more sensitive and accurate and requires less sample; also the fluorescence is more stable than is the color obtained in the colorimetric method.

### 2.3.2. Fluorescent Assay of Total Serum Cholesterol, with Use of Gas-Liquid Chromatography to Study Saponification Efficiency (302)

Elevitch (303) has summarized the fluorometric methods for measuring cholesterol in microscale amounts of serum and other biological fluids.

In 1872, Salkowski (304) described the series of colors and the fluorescence seen when concentrated sulfuric acid is added to an equal volume of a solution of cholesterol in chloroform. Later workers (305–308) emphasized that anhydrous conditions, control of temperature, and the use of a consistent fluorescent development time are important in the Salkowski reaction. In 1943, Merkelbach (307) reported that the fluorescence of cholesterol in sulfuric acid, as recorded on photographic plates, fell in the 470–650 nm region of the spectrum. He suggested that such broad fluorescence was attributable to the variable formation of different isomeric compounds. Using newer, commercially available fluorometric instrumentation, Majeski et al. (302) found that the Salkowski reaction product fluoresces strongly between 430 and 530 nm on excitation at 415 nm. They also found that methylene chloride is a better solvent than chloroform to use in the Salkowski reaction: it provides a lower solvent blank, more readily dissolve the cholesterol standards, and is more stable to attack by base than is chloroform.

The present work by Majeski et al. describes a fluorescent determination of total cholesterol in serum for which the accuracy and precision are comparable to those for the standard Abell-Kendall method (309–311), a method of generally accepted accuracy. By the use of quality reagents and the rigorous exclusion of water, the strong fluorophor (Fig. 4.44) that develops on reacting concentrated sulfuric acid with cholesterol can be used to quantitatively determine the total cholesterol in serum. Gas-liquid chromatography was used to monitor the extent of saponification of the cholesterol esters because they have fluorescent efficiencies that differ from the efficiency of free cholesterol. Sodium methoxide in methanol/methylene chloride (1/3 by volume) was shown by gas-liquid chromatography to saponify very effectively the cholesterol esters in serum.

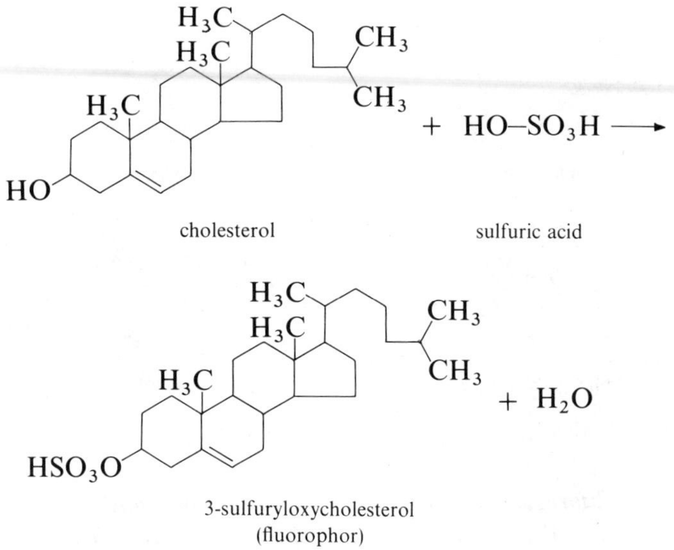

**Figure 4.44.** Derivatization of cholesterol with sulfuric acid.

### 2.3.3. New Fluorometric Analysis for Mandelic and Phenylglyoxylic Acids in Urine as an Index of Styrene Exposure (312)

About 70–80% of the total production of styrene is used in the production of plastic polymers and about 15% in the manufacture of synthetic rubber. Most of the styrene taken up by workers is metabolized; only about 1% is expired unchanged (313). Only mandelic acid (MA) and phenylglyoxylic acid (PGA) are established as being major metabolites of styrene in human urine, and the usefulness of urinary MA and PGA as biological measures of current styrene exposure has been well documented (313–318). So far either colorimetric (313) or gas-chromatographic (314–318) methods have been used to estimate these two acids in urine.

When aromatic compounds are dissolved in concentrated sulfuric acid, various fluorescent protonated derivatives, the hydroxycarbonium ion (Fig. 4.45), are formed (319). Chakrabarti (312) applied this principle to a new fluorometric method for determination of MA and PGA in urine, the fluorometry being preceded by extraction into ether and thin-layer chromatography. The chromatographically separated acids were quantitated after conversion to stable higher fluorescent derivatives by treatment with concentrated sulfuric acid. This method has proved to be more precise, accurate, and reproducible than the existing colorimetric method.

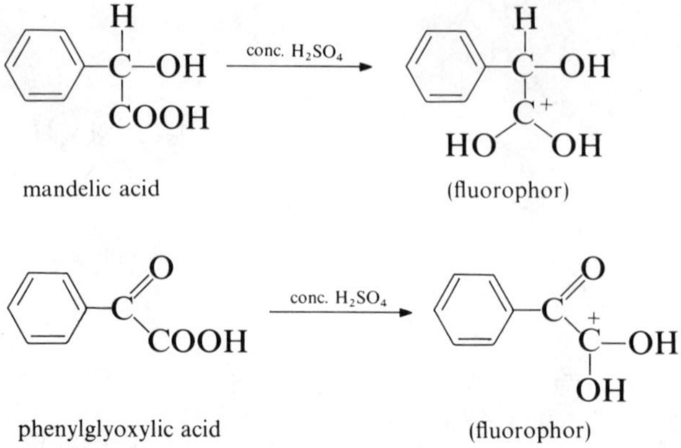

## Figure 4.45. Derivatization of mandelic and phenylglyoxylic acids with sulfuric acid.

### 2.3.4. Fluorescence Methods in the Diagnosis and Management of Diseases of Tetrapyrrole Metabolism (320)

Under appropriate conditions fluorescent porphyrins and bilirubin (Fig. 4.46) present in blood and other body fluids can be examined spectrofluorometrically without prior extraction. The uses of such direct fluorescence spectroscopy of porphyrins and bilirubin in studies and diagnoses of diseases associated with abnormal or impaired heme synthesis and metabolism were discussed by Lamola (320). The method of "front-face" fluorometry, which allows quantitive assays of fluorescent porphyrins and bilirubin in small, undiluted blood specimens, was described.

### 2.3.5. Sensitive Fluorometry of Heat-Stable Alkaline Phosphatase (Regan Enzyme) Activity in Serum from Smokers and Nonsmokers (321)

Maslow et al. (321) have developed a simple, sensitive enzymatic assay involving the fluorogenic substrate naphthol AS-MX phosphate [(3-hydroxy-2-naphthoic acid-2,4-dimethylanilide) phosphate; Fig. 4.47]

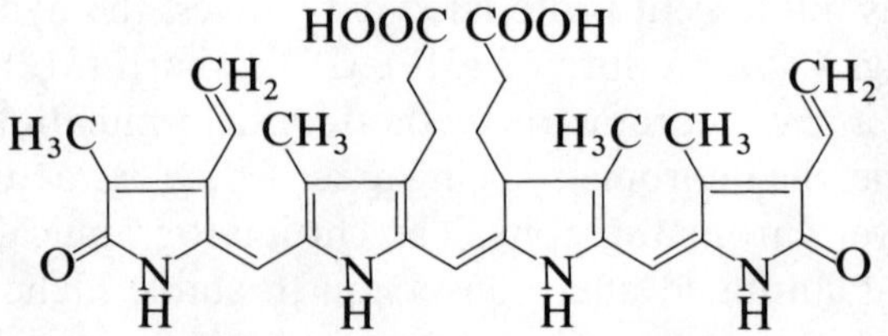

**Figure 4.46.** Structure of bilirubin.

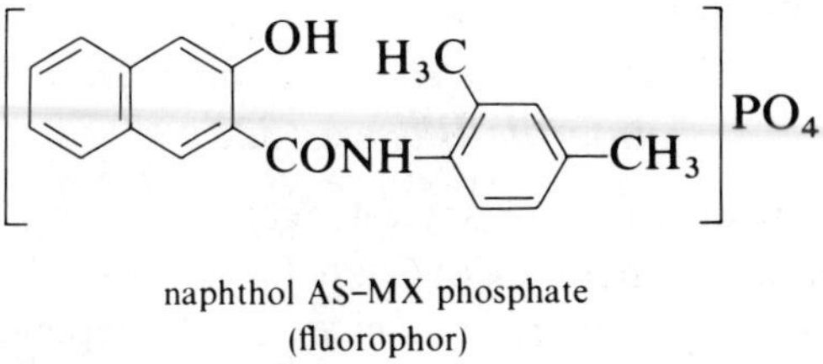

naphthol AS–MX phosphate
(fluorophor)

**Figure 4.47.** Fluorogenic substrate for the fluorescent determination of HSAP.

to measure heat-stable alkaline phosphatase (HSAP; EC 3.1.3.1), the Regan isoenzyme, in human serum.

The mean for 51 nonsmokers was 0.068 (S.D. = 0.037) arb. units $dm^{-3}$; for 25 smokers it was 0.440 (S.D. = 0.360) arb. units $dm^{-3}$. Activity of this isoenzyme in smokers was as much as 10-fold the upper normal limit for nonsmokers. Activation of this tumor marker by smoking has not received attention hitherto. The present authors conclude that a truly normal range can be established only among nonsmokers.

## 2.4. Obstetrics and Gynecology

### 2.4.1. *New Rapid Assay of Estrogens in Pregnancy Urine Using the Substrate Native Fluorescence (322)*

A fluorometric method is described by Bramhall and Britten (322) for the quantitative determination of $> 2 \,\mu g \, cm^{-3}$ (7 mmol $m^{-3}$) of free and conjugated estrogens in 24-h pregnancy urine. The estrogens were precipitated with $(NH_4)_2SO_4$ and freed from nonestrogenic compounds by solvent extraction. The conjugated estrogens were hydrolyzed by a $\beta$-glucuronidase from *Escherichia coli*, the total free estrogens were extracted into $Et_2O$, and their fluorescence intensity at 310 nm in this solvent was determined. This method, which measures estradiol and estriol levels (Fig. 4.48), can be applied routinely to monitor fetoplacental function in pregnancy.

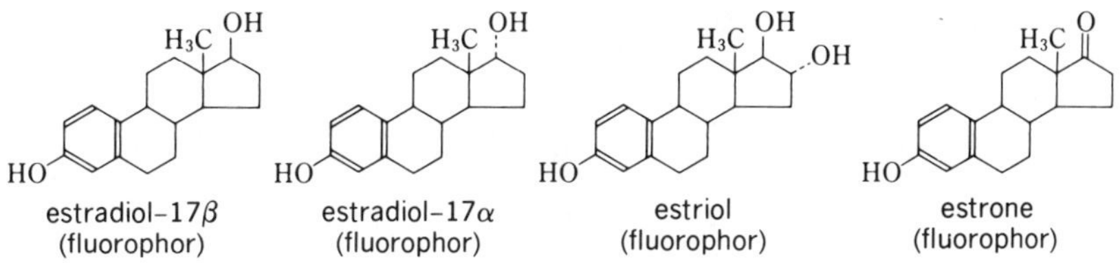

estradiol-17$\beta$
(fluorophor)

estradiol-17$\alpha$
(fluorophor)

estriol
(fluorophor)

estrone
(fluorophor)

**Figure 4.48.** Native fluorophors of estrogens.

### 2.4.2. A Semiautomated Method for the Determination of Estrogens in Early Morning Urine Specimens from Normal and Infertile Women (323)

The measurement of total urinary estrogens has long been regarded as useful in the assessment of ovarian function (324), although the reference method (325) is work intensive, is time dependent, and relies on a complete 24-h collection of urine. Beastall and McVeigh (323) describe a partial automation (326, 327) for the routine fluorometric

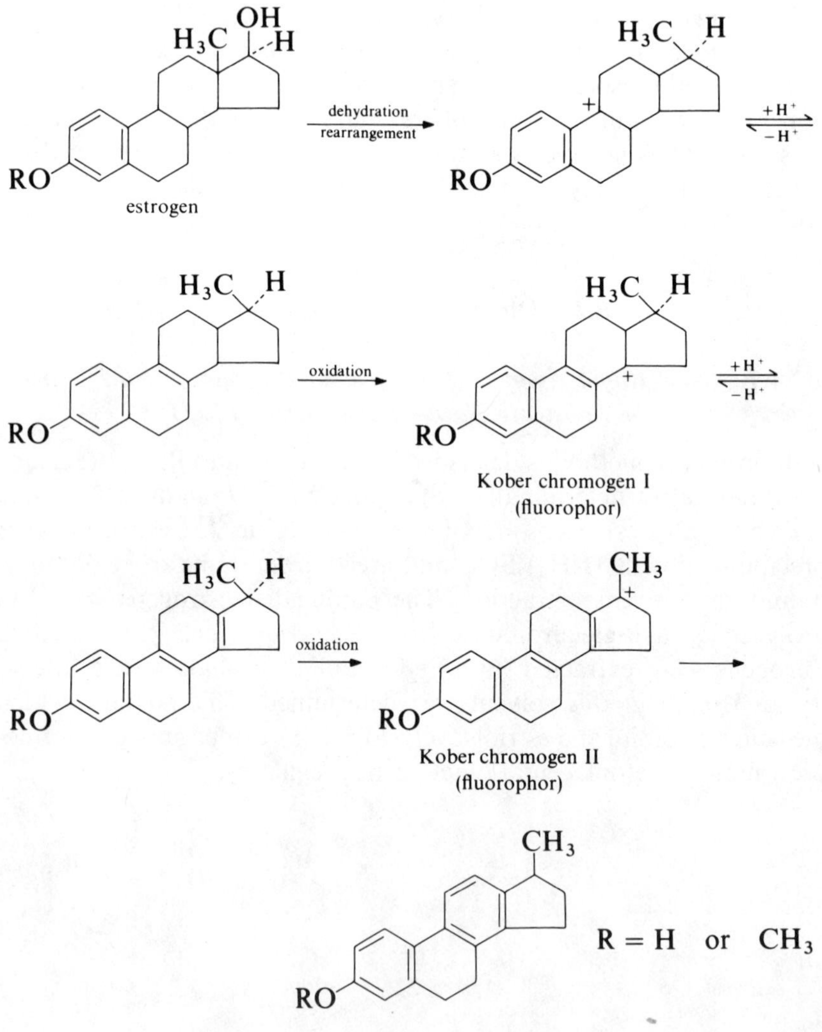

**Figure 4.49.** Mechanism of Kober's reaction.

determination of estrogens, and examine data obtained from the measurement of the urinary estrogen/creatinine (E/C) ratio in early morning urine specimens (EMU). Comparative studies reveal that the estrogen content of an early morning urine specimen is a clinically useful parameter in the investigation of female infertility.

*Procedure:* Aliquots of nonpregnancy urine (2 cm$^3$) are hydrolyzed in a small sterilizer, and partially purified estrogen extracts are obtained. The dried extracts are solubilized in 5% ethanol/water (2 cm$^3$) and sampled at a rate of 20/h by an Autoanalyzer Sampler. The urine extract is mixed with Kober reagent (189 mol m$^{-3}$ quinol in 66% sulfuric acid), and the chromogen developed (see Fig. 4.49) in a 12 m × 1.6 mm glass coil maintained at a temperature of 125°C. An extraction with Ittrich reagent (306 mol m$^{-3}$ trichloroacetic acid in chloroform) is effected, and the fluorescence of the extract determined with a fluorometer.

An example of the experimental results is shown in Fig. 4.50.

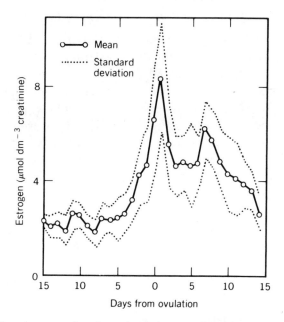

**Figure 4.50.** Estrogen excretion throughout six normal cycles. Reprinted from G. H. Beastall and S. McVeigh, *Clin. Chim. Acta*, **70**, 343 (1976), Elsevier Science Publishers, The Netherlands.

### 2.4.3. Semiautomated Fluorometric Method for the Determination of Total Estrogens in Pregnancy Urine (328)

Frye et al. (328) report a continuous-flow fluorometric method for total urinary estrogens that involves the Kober reaction, with extraction of the reaction product into dichloroethane containing trichloroacetic acid as described by Hahnel and Jones (329). The dichloroethane extraction gives greater stability to the Kober color and sharper separation of the aqueous and organic phases. The analytical system is adjusted to give maximum response to estriol 16α,β-D-glucuronide, the principal estrogen conjugate in urine from late pregnancy, when calibrated with estriol standards. An initial 20-fold dilution of the sample with water increased analytical recovery of estriol conjugates from urine while maintaining adequate fluorescent response. Glucose interference was reduced by dilution and eliminated by treatment with sodium borohydride. Urinary protein up to 20 g dm$^{-3}$ did not interfere. Total estriol, as determined by gas-liquid chromatography, comprises about 70% of total urinary estrogens in late pregnancy, as measured by the present continuous-flow fluorometric method.

### 2.4.4. Evaluation of an Aqueous Fluorometric Continuous-Flow Method for Measurement of Total Urinary Estrogens (330)

Measurement of urinary estrogen excretion is used extensively to monitor fetal welfare *in utero*, allowing the fetus at risk to be identified. Hammond et al. (330) evaluated a direct all-aqueous fluorometric method adapted to a continuous-flow system, using a spectrofluorophotometer for detection. In this method, which is based on a modified Kober reaction (see Fig. 4.49), NaBH$_4$ is used to eliminate the known negative interference of glucose. These workers conclude that this automated fluorometric method is accurate, precise, and inexpensive, and therefore is the method of choice for measuring urinary estrogens.

## 2.5. Other Clinical Fields

### 2.5.1. Inherent Fingerprint Luminescence—Detection by Laser (331)

The currently used methods of fingerprint detection may be classified into two categories: those that depend on the adherence of inert materials to fingerprint residues (powder methods), and those that rely on chemical interaction of a detection reagent with specific components of the latent print (e.g., the ninhydrin method) (332). Both types—in fact, all conventional fingerprint detection methods—require chemical or physical treatment of the exhibit under examination.

Exploitation of an intrinsic property of components present in latent fingerprints, such that development could be accomplished without "staining" the exhibit under scrutiny, would be useful. Such a procedure would circumvent possible deleterious effects of the fingerprint development step on other exhibit examination procedures, whether for fingerprint detection, analysis for blood, or some other purpose.

Palmar sweat contains a variety of compounds, among them amino acids, lipids, and vitamins (333). Some of these compounds show native fluorescence. For example, riboflavin and pyridoxin (Fig. 4.51), present in palmar sweat, show fluorescence at 565 and 400 nm, respectively (334). Since such compounds are present in fingerprint deposits in rather small quantities, however, fingerprints do not show discernible luminescence under normal illumination conditions. However, spectroscopists have known for some time that fingerprints can luminesce under laser illumination.

Dalrymple et al. (331) deal with an initial exploration of the utilization of lasers in fingerprint detection. The fingerprint detection procedure consisted essentially of laser illumination of the exhibit under scrutiny and photography (or direct viewing) of the laser-induced fingerprint luminescence. Figure 4.52 illustrates the experimental configuration. Exhibits under investigation were illuminated with the 514.5-nm line of a continuous-wave (CW) argon-ion laser. The laser power was 1.5 W, and the laser beam was expanded to illuminate an area of 10 in.$^2$ (64.5 cm$^2$).

### 2.5.2. A New Dual-Staining Technique for Simultaneous-Flow Cytometric DNA Analysis of Living and Dead Cells (335)

Stöhr and Vogt-Schaden (335) accomplished living-dead cell discrimination by dual staining of DNA from viable cells of a Walker ascites

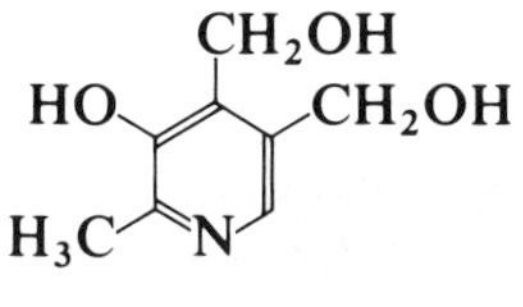

(fluorophor)

Figure 4.51. Structure of pyridoxine.

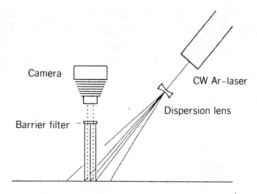

**Figure 4.52.** Schematic diagram of laser detection setup (dotted lines denote fingerprint luminescence.) [From B. E. Dalrymple, J. M. Duff, and E. R. Menzel, *J. Forensic Sci.*, **22**, 106 (1977). Copyright ASTM. Reprinted with permission.]

**Figure 4.53.** Structure of benzimidazole as a fluorescent staining.

tumor by using HOE 33662 (a benzimidazole compound from Hoechst; Fig. 4.53) and propidium iodide (PI), followed by computer-assisted flow cytometry. Viability tests were carried out using fluorescein diacetate (FDA). Under UV illumination, the living cells (FAD-pos.) exhibited blue fluorescence due to accumulation of HOE 33662 in the chromation, whereas the dead cells (FAD-neg.) exhibited red fluorescence due to penetration of PI into the interior of the dead cells and transfer of the radiation energy from HOE 33662 to PI. In both viable and dead cells, the fluorescence intensity was proportional to the DNA content. Under green illumination, which selectively causes excitation of PI, only the dead cells showed red fluorescence. The two-dimensional computer-drawn frequency distribution of Walker ascites tumor cells, as well as of other tumor cell lines, showed definite separation of living and dead cells and of an intermediate area of dying cells (Fig. 4.54). The biomedical applications of the method are also discussed.

### 2.5.3. Continuous Measurement of Concentrations of Alcohol Using a Fluorescence-Photometric Enzymatic Method (336)

Using fluorescence optical techniques instead of electrodes (337), Völkl et al. (336) have developed a method for determining the

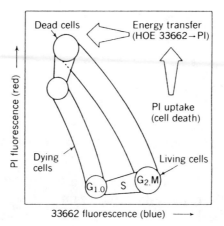

**Figure 4.54.** Schematic diagram of the dual fluorescence differentiation of living, dying, and dead cells in an experimental ascites tumour. The red fluorescence of PI as derived from energy transfer from 33662 to PI (vertical axis) is plotted versus the total blue fluorescence of 33662.

concentrations of ethanol in biological fluids. The determination is based on the measurement of the oxygen consumption caused by oxidation of ethanol. The reaction is catalyzed by alcohol oxidase (AOD, EC 1.1.3.13). The aldehyde formed must be promptly converted to physiologically indifferent products.

The enzymes are incorporated in a gel layer. Depending on its concentration, alcohol diffuses into the layer and reacts with oxygen (1 mol dm$^{-3}$ alcohol consumes 1 mol dm$^{-3}$ oxygen). An oxygen pressure difference develops across the gel layer that is proportional to the alcohol concentration, provided that the following conditions are met:

1. Oxygen is available in excess.
2. The enzyme concentration is in a range where velocity is proportional to the substrate concentration.
3. The reaction products are in diffusion equilibrium with the medium to be measured.

Under these conditions, $p_{O_2}$ decreases across the enzyme layer in a cosh function. Pyrene butyric acid (Fig. 4.55) was used as an oxygen indicator. The oxygen sensitivity of the fluorescence intensity of the indicator is given by Eq. (4.5), according to Vaughan and Weber (338):

$$\frac{I_0}{I} = 1 + Kp_{O_2} \qquad (4.5)$$

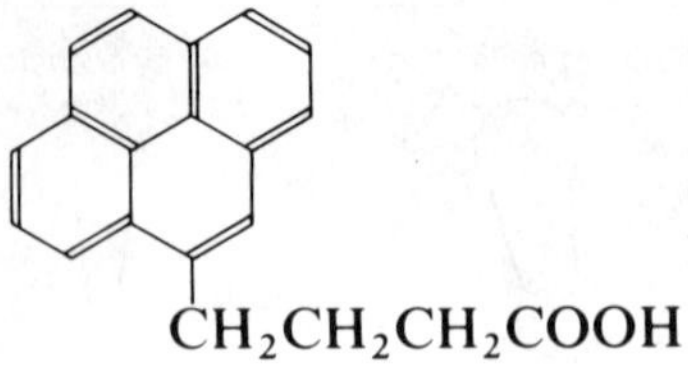

CH₂CH₂CH₂COOH

(fluorophor)

**Figure 4.55.** Structure of pyrene butyric acid as an oxygen indicator.

where $I_0$ is the relative fluorescence intensity at oxygen partial pressure 0 atm., $I$ is the relative fluorescence intensity at the actual oxygen partial pressure, and the constant, $K$, is determined by calibration with different oxygen pressures.

When plotting the quotient $I_0/I$ against the alcohol concentration while $p_{O_2}$ in the medium on the front side of the enzyme layer is constant, the present authors obtain a linear calibration curve:

$$\frac{I_0}{I} = 1 + K p_{O_2} - K' \text{[alcohol]} \qquad (4.6)$$

where $K'$ denotes the parameter specific for the enzyme layer.

AOD is satisfactorily specific to determine ethanol in biological fluids, because polyhydric alcohols such as glycerine are not converted (339) and alcohols with a measurable reaction rate, except ethanol, are not present in biological fluids. The reaction products, acetaldehyde and hydrogen peroxide, are converted to indifferent products such as acetate and water in a subsequent reaction.

## REFERENCES

1. S. Udenfriend and P. Zaltzman, *Anal. Biochem.*, **3**, 49 (1962).
2. F. W. J. Teale and G. Weber, *Biochem. J.*, **65**, 476 (1957).
3. D. G. Kaiser, W. F. Liggett, and A. A. Forist, *J. Pharm. Sci.*, **65**, 1767 (1976).
4. E. Hohaus, *Fresenius Z. Anal. Chem.*, **310**, 70 (1982).
5. T. K. Hwang et al., *Anal. Chim. Acta*, **99**, 305 (1978).
6. R. Wintersteiger et al., *Fresenius Z. Anal. Chem.*, **312**, 455 (1982).
7. A. DeLeenheer, J. E. Sinsheimer, and J. H. Burckhalter, *J. Pharm. Sci.*, **62**, 1370 (1973).
8. H. Nakamura et al., *Anal. Chem.*, **54**, 2482 (1982).

# REFERENCES

9. R. W. Frei et al., *HRC CC, J. High Resolut. Chromatogr. Commun.*, **2**, 11 (1979).
10. U. A. T. Brinkman et al., *Anal. Chem. Symp. Ser.*, **3**, 247 (1980).
11. S. Udenfriend et al., *Science*, **178**, 871 (1972).
12. M. Weigele et al., *J. Am. Chem. Soc.*, **94**, 5927 (1972).
13. P. M. Froehlich and T. D. Cunningham, *Anal. Chim. Acta*, **84**, 427 (1976).
14. J. Kusnir, *Chem. Listy*, **71**, 1046 (1977).
15. J. V. Castell, M. Cervera, and R. Marco, *Anal. Biochem.*, **99**, 379 (1979).
16. B. A. Tomkins, V. H. Ostrum, and C. H. Ho, *Anal. Lett.*, **13**, 589 (1980).
17. H. Khalaf and M. Rimpler, *Hoppe-Seyler's Z. Physiol. Chem.*, **358**, 505 (1977).
18. N. Seiler and M. Wiechmann, *Z. Anal. Chem.*, **220**, 109 (1966).
19. B. Leonard and N. N. Osborne, *Res. Methods Neurochem.*, **3**, 443 (1975).
20. P. L. Felgner and J. E. Wilson, *Anal. Biochem.*, **80**, 601 (1977).
21. B. A. Davis, *J. Chromatogr.*, **151**, 252 (1978).
22. N. Seiler, T. Schmidt-Glenewinkel, and H. H. Schneider, *J. Chromatogr.*, **84**, 95 (1973).
23. P. B. Ghosh and M. W. Whitehouse, *Biochem. J.*, **108**, 155 (1968).
24. K. Samejima et al., *Anal. Biochem.*, **42**, 222 (1971).
25. N. Lustenberger, H. W. Lange, and K. Hempel, *Angew. Chem.*, **84**, 255 (1972).
26. G. Talsky and C. Gramsch, *Z. Anal. Chem.*, **282**, 451 (1976).
27. S. Kirschenbaum and M. D. Glantz, *Mikrochim. Acta*, **1**, 289 (1976).
28. J. T. Stewart and D. M. Lotti, *Anal. Chim. Acta*, **41**, 178 (1968).
29. M. Roth, *Anal. Chem.*, **43**, 880 (1971).
30. M. Roth and A. Hampai, *J. Chromatogr.*, **83**, 353 (1973).
31. S. S. Simons, Jr., D. F. Johnson, *Anal. Biochem.*, **82**, 250 (1977).
32. R. Hakanson, A. L. Roennberg, and K. Sjoelund, *Anal. Biochem.*, **59**, 98 (1974).
33. E. Mendez and J. G. Gavilanes, *Anal. Biochem.*, **72**, 473 (1976).
34. G. J. Schmidt et al., *Chromatogr. Newsl.*, **5**, 33 (1977).
35. J. A. F., DeSilva and N. Strojny, *Anal. Chem.*, **47**, 714 (1975).
36. J. E. Sinsheimer et al., *J. Pharm. Sci.*, **57**, 1938 (1968).
37. L. Brand and J. R. Gohlke, *Ann. Rev. Biochem.*, **41**, 843 (1972).
38. V. R. Villanueva et al., *J. Chromatogr.*, **139**, 381 (1977).
39. G. Milano et al., *J. Clin. Chem. Clin. Biochem.*, **18**, 157 (1980).
40. G. Schwedt, *Z. Anal. Chem.*, **287**, 152 (1977).
41. A. E. Ciarlone, *Microchem. J.*, **21**, 349 (1976).
42. A. E. Ciarlone, *Microchem. J.*, **23**, 9 (1978).

43. J. F. Lawrence and R. W. Frei, *Anal. Chem.*, **44**, 2046 (1972).
44. R. J. Sturgeon and S. G. Schulman, *Anal. Chim. Acta*, **75**, 225 (1975).
45. H. J. Klimisch and L. Stadler, *J. Chromatogr.*, **90**, 141 (1974).
46. H. J. Klimisch and L. Stadler, *J. Chromatogr.*, **90**, 223 (1974).
47. M. Maeda, T. Kinoshita, and A. Tsuji, *Anal. Biochem.*, **38**, 121 (1970).
48. M. Maeda and A. Tsuji, *Anal. Biochem.*, **52**, 555 (1973).
49. M. Pesez and J. Bartos, *Talanta*, **16**, 331 (1969).
50. M. Pesez and J. Bartos, *Ann. Pharm. Franc.*, **27**, 161 (1969).
51. J. Bartos, *Talanta*, **21**, 1303 (1974).
52. L. J. Dombrowski and E. L. Pratt, *Anal. Chem.*, **43**, 1042 (1971).
53. H. Taniguchi et al., *Chem. Pharm. Bull.*, **29**, 784 (1981).
54. W. H. Harrison, *Arch. Biochem. Biophys.*, **101**, 116 (1963).
55. H. Weil-Malherbe and J. B. Bigelow, *Anal. Biochem.*, **22**, 321 (1968).
56. H. Wisser and E. Knoll, *Z. Klin. Chem. Klin. Biochem.*, **11**, 3 (1973).
57. H. Weil-Malherbe, *Biochem. Biophys. Acta*, **40**, 351 (1960).
58. S. M. Hess and S. Udenfriend, *J. Pharmacol. Exptl. Therap.*, **127**, 175 (1959).
59. E. Hohaus, *Bunseki Kagaku*, **33**, E 55 (1984).
60. H. Khalaf and M. Rimpler, *Fresenius Z. Anal. Chem.*, **302**, 203 (1980).
61. H. Nakamura and Z. Tamura, *Anal. Chem.*, **52**, 2087 (1980).
62. A. Himuro et al., *Anal. Chim. Acta*, **147**, 317 (1983).
63. L. L. Dent et al., *Anal. Lett.*, **14**, 1031 (1981).
64. R. P. Andrews and N. A. Baldar, *Sci. Tools*, **30**, 8 (1983).
65. T. Matsumoto et al., *Clin. Chim. Acta*, **112**, 141 (1981).
66. C. Beyer and A. Van den Ende, *Clin. Chim. Acta*, **129**, 211 (1983).
67. D. Lindvall et al., *Methods Neurobiol.*, **2**, 365 (1981).
68. S. Udenfriend, *Fluorescnence Assay in Biology and Medicine*, Vol. 1, Academic Press, New York, 1962.
69. S. Udenfriend, *Fluorescence Assay in Biology and Medicine*, Vol. 2, Academic Press, New York, 1969.
70. G. Schwedt, *Anal. Chim. Acta*, **81**, 361 (1976).
71. G. Schwedt, *J. Chromatogr.*, **118**, 429 (1976).
72. C. Reinhold and H. G. Hartwig, *Brain Res. Bull.*, **9**, 97 (1982).
73. S. Kamata et al., *J. Chromatogr.*, **231**, 291 (1982).
74. F. W. J. Teale and G. Weber, *Biochem. J.*, **65**, 476 (1957).
75. P. M. Froehlich and M. Yeats, *Anal. Chim. Acta*, **87**, 185 (1976).
76. W. Altekar, *Biopolymers*, **16**, 341 (1977).
77. E. Trepman and R. F. Chen, *Arch. Biochem. Biophys.*, **204**, 524 (1980).
78. D. J. Shute, *Med. Lab. Sci.*, **37**, 173 (1980).

# REFERENCES

79. A. J. Thomas, *Amino Acid Anal. (Symp.)*, **1979**, 37 (1981).
80. Y. Sugimura and Y. Suzuki, *Pap. Meteorol. Geophys.*, **33**, 269 (1983).
81. K. Yokotsuka and T. Kushida, *J. Ferment. Technol.*, **61**, 1 (1983).
82. A. B. Bleecker and J. T. Romeo, *Anal. Biochem.*, **121**, 295 (1982).
83. A. S. Kimes and M. K. Shellenberger, *Pharmacol. Biochem. Behav.*, **18**, 943 (1983).
84. P. Boehlen and R. Schroeder, *Anal. Biochem.*, **126**, 144 (1982).
85. E. Mendez, *Anal. Biochem.*, **127**, 55 (1982).
86. T. Iqbal et al., *J. Chem. Soc. Pak.*, **2**, 141 (1980).
87. K. Imai and Y. Watanabe, *Anal. Chim. Acta*, **130**, 377 (1981).
88. H. Khalaf and M. Rimpler, *Fresenius Z. Anal. Chem.*, **302**, 203 (1980).
89. H. N. Singh and W. L. Hinze, *Analyst (London)*, **107**, 1073 (1982).
90. M. Snejdarkova and M. Otto, *Z. Chem.*, **21**, 229 (1981).
91. M. Barcelon et al., *J. Chromatogr.*, **260**, 147 (1983).
92. H. Khalaf et al., *Fresenius Z. Anal. Chem.*, **302**, 415 (1980).
93. K. Murayama and T. Kinoshita, *Chem. Pharm. Bull.*, **28**, 1925 (1980).
94. K. Murayama and T. Kinoshita, *J. Chromatogr.*, **205**, 349 (1981).
95. K. Murayama and T. Kinoshita, *Anal. Lett.*, **15**, 123 (1982).
96. G. G. Guiltbault, *Practical Fluorescence—Theory, Methods and Techniques*, M. Dekker, New York, 1973.
97. R. A. Chalmers and G. A. Wadds, *Analyst*, **95**, 234 (1970).
98. M. Montagu-Bourin et al., *Planta Med.*, **38**, 50 (1980).
99. T. Gürkan, *Microchim. Acta*, **1**, 173 (1976).
100. D. Naik et al., *Anal. Chim. Acta*, **74**, 29 (1975).
101. S. M. Hassan, *J. Pharm. Belg.*, **38**, 305 (1983).
102. A. D. Thomas, *Talanta*, **22**, 865 (1975).
103. M. Montagu et al., *Talanta*, **28**, 709 (1981).
104. J. Hetherington et al., *J. Photochem.*, **20**, 367 (1982).
105. C. W. McLeod and T. S. West, *Analyst*, **107** (1270), 1 (1982).
106. N. V. R. Rao and S. N. Tandon, *Anal. Lett.*, **311**, 477 (1978).
107. A. Szabo and E. M. Karacsony, *J. Chromatogr.*, **193**, 500 (1980).
108. W. A. Hagins and W. H. Jennings, *Discussions Faraday Soc.*, **27**, 180 (1959).
109. J. J. Aaron and J. D. Winefordner, *Talanta*, **19**, 21 (1972).
110. S. Udenfriend, *Fluorescence Assay in Biology and Medicine*, Vol. 1, Academic Press, New York, 1962.
111. D. E. Duggan et al., *Arch. Biochem. Biophys.*, **68**, 7 (1957).
112. J. Kahan, *Scand. J. Clin. Lab. Invest.*, **18**, 679 (1966).
113. B. D. Drujan et al., *Anal. Biochem.*, **23**, 44 (1968).

114. O. A. Bessey et al., *J. Biol. Chem.*, **180**, 755 (1949).
115. G. Weber, *Biochem. J.*, **47**, 114 (1950).
116. M. A. Ryan and J. D. Ingle, Jr., *Talanta*, **28**, 225 (1981).
117. D. B. Coursin and V. C. Brown, *Proc. Soc. Expm. Biol. Med.*, **98**, 315 (1958).
118. J. W. Bridges et al., *Biochem. J.*, **98**, 451 (1966).
119. R. F. Chen, *Science*, **150**, 1593 (1965).
120. S. Udenfriend, *Fluorescence Assay in Biology and Medicine*, Vol. 2, Academic Press, New York, 1969.
121. D. E. Duggan, *Arch. Biochem. Biophys.*, **84**, 116 (1959).
122. W. E. Ohnesorge and L. B. Rodgers, *Anal. Chem.*, **28**, 1017 (1956).
123. K. Yagi et al., *Vitamins (Kyoto)*, **9**, 391 (1955).
124. H. B. Burch et al., *J. Biol. Chem.*, **198**, 477 (1952).
125. K. L. Muiruri et al., *Am. J. Clin. Nutr.*, **27**, 837 (1974).
126. O. Pelletier and R. Madere, *Clin. Chem.*, **18**, 937 (1972).
127. B. Karlberg and S. Thelander, *Anal. Chim. Acta*, **114**, 129 (1980).
128. J. V. Scudi, *Science*, **103**, 567 (1946).
129. T. E. Friedemann and E. I. Frazier, *Science*, **102**, 97 (1945).
130. J. W. Huff, *J. Biol. Chem.*, **167**, 151 (1947).
131. H. B. Burch et al., *J. Biol. Chem.*, **212**, 897 (1955).
132. M. J. Deutsch and C. E. Weeks, *J. Assoc. Pffic. Agr. Chemists*, **48**, 1248 (1965).
133. S. W. Jones et al., *Am. Chem. Soc. (Abs. 138th Meeting) New York Sept.*, 1960, p. 60 c (1960).
134. P. S. Chen et al., *Anal. Biochem.*, **8**, 34 (1964).
135. O. Warburg and W. Christian, *Biochem. Z.*, **266**, 377 (1933).
136. K. Yagi, *Bull. Soc. Chim. Fr.*, 1543 (1957).
137. V. Allfrey et al., *J. Biol. Chem.*, **178**, 465 (1949).
138. S. Udenfriend, *Fluorescence Assay in Biology and Medicine*, Vol. 1, Academic Press, New York, 1962.
139. J. Bramhall and A. Z. Britten, *Clin. Chim. Acta*, **68**, 203 (1976).
140. S. Kober, *Biochem. Z.*, **239**, 209 (1931).
141. G. Z. Ittrich, *Z. Physiol. Chem.*, **312**, 1 (1958).
142. G. Z. Ittrich, *Acta Endocrinol.*, **35**, 34 (1960).
143. S. Udenfriend, *Fluorescence Assay in Biology and Medicine*, Vol. 2, Academic Press, New York, 1969.
144. L. Lee and R. Haehnel, *Clin. Chem.*, **17**, 1194 (1971).
145. K. H. Outch et al., *Clin. Chim. Acta*, **40**, 377 (1972).
146. G. F. Taylor and N. G. Carter, *Pathology*, **13**, 97 (1981).
147. T. Nakao and Y. Aizawa, *Endocrinol. Japan*, **3**, 92 (1956).

148. E. Epstein and B. Zak, *Clin. Chem.*, **9**, 70 (1963).
149. A. Smoczkiewiczowa and R. Sioda, *J. Pharm. Pharmacol.*, **15**, 486 (1963).
150. R. W. Albers and O. H. Lowry, *Anal. Chem.*, **27**, 1829 (1955).
151. D. B. McDougal, Jr. and H. S. Farmer, *J. Lab. Clin. Med.*, **50**, 485 (1957).
152. G. J. Koval, *J. Lipid Res.*, **2**, 419 (1961).
153. E. B. Solow and L. W. Freeman, *Clin. Chem.*, **16**, 472 (1970).
154. J. C. Fruchart et al., *Anals. Biol. Clin.*, **33**, 453 (1975).
155. M. L. Sweat, *Anal. Chem.*, **26**, 773 (1954).
156. H. Kalant, *Biochem. J.*, **69**, 79 (1958).
157. J. W. Goldzieher and P. H. Besch, *Anal. Chem.*, **30**, 962 (1958).
158. N. Zenker and D. E. Bernstein, *J. Biol. Chem.*, **231**, 695 (1958).
159. R. H. Silber, *Clin. Chem.*, **4**, 278 (1958).
160. H. Braunsberg and V. H. T. James, *J. Endocrinol.*, **25**, 309 (1962).
161. I. E. Bush and A. A. Sandberg, *J. Biol. Chem.*, **205**, 783 (1953).
162. P. J. Ayres et al., *Biochem. J.*, **65**, 639 (1957).
163. P. J. Ayres et al., *Biochem. J.*, **65**, 647 (1957).
164. D. Abelson and P. K. Bondry, *Arch. Biochem. Biophys.*, **57**, 208 (1955).
165. J. Koolman, *Insect Biochem.*, **10**, 381 (1980).
166. R. J. Spooner, P. A. Toseland, and D. M. Goldberg, *Clin. Chim. Acta*, **65**, 47 (1975).
167. M. Rubin and L. Knott, *Clin. Chim. Acta*, **18**, 409 (1967).
168. H. Poiger and Ch. Schlatter, *Analyst*, **101**, 808 (1976).
169. L. R. Peterson, P. Hamernyik, T. D. Bird, and R. F. Labbé, *Clin. Chem.*, **22**, 1835 (1976).
170. R. Beke, G. A. De Weerdt, J. Parijs, and F. Barbier, *Clin. Chim. Acta*, **71**, 27 (1976).
171. G. W. Hepner, A. F. Hofmann, J. R. Malagelada et al., *Gastroenterology*, **66**, 556 (1974).
172. J. Sjövall, *Clin. Chim. Acta*, **4**, 652 (1959).
173. D. Panveliwalla, B. Lewis, I. D. P. Wootton, and S. Tabaqchali, *J. Clin. Pathol.*, **23**, 309 (1970).
174. D. D. Jones, *Clin. Chim. Acta*, **19**, 57 (1968).
175. D. H. Sandberg, J. Sjövall, K. Sjövall, and D. A. Turner, *J. Lipid Res.*, **6**, 182 (1965).
176. G. P. Van Berge Henegouwen, A. Ruben, and K. H. Brandt, *Clin. Chim. Acta*, **54**, 249 (1974).
177. N. Kaplowitz and N. B. Javitt, *J. Lipid Res.*, **14**, 224 (1973).
178. T. Sumi, Y. Umeda, Y. Kishi, K. Takahashi, and F. Kamimoto, *Clin. Chim. Acta*, **73**, 233 (1976).

179. A. Ågren, S. E. Brolin, and S. Hjerten, *Biochem. Biophys. Acta*, **500**, 103 (1977).
180. G. A. Mason, G. K. Summer, H. H. Dutton, and R. C. Schwaner, Jr., *Clin. Chem.*, **23**, 917 (1977).
181. T. Osuga, K. Mitamura, F. Mashige, and K. Imai, *Clin. Chim. Acta*, **75**, 81 (1977).
182. G. G. Guilbault and D. N. Kramer, *Anal. Chem.*, **37**, 1219 (1965).
183. F. Mashige, K. Imai, and T. Osuga, *Clin. Chim. Acta*, **70**, 79 (1976).
184. G. Curzon and B. D. Kantamaneni, *Clin. Chim. Acta*, **76**, 289 (1977).
185. V. P. Dole, *J. Clin. Invest.*, **35**, 150 (1956).
186. F. G. Soloni and L. C. Sardina, *Clin. Chem.*, **19**, 419 (1973).
187. J. Williamson and T. J. Scott-Finnigan, *Clin Chim. Acta*, **57**, 175 (1974).
188. V. P. Dole, *J. Clin. Invest.*, **35**, 150 (1956).
189. M. G. Brunett, N. Vigneault, and G. Daman, *Anal. Biochem.*, **86**, 229 (1978).
190. R. D. Bell and E. A. Doisy, *J. Biol. Chem.*, **44**, 55 (1920).
191. J. F. Boudry, U. Troehler, M. Touabi et al., *Clin. Sci. Mol. Med.*, **48**, 475 (1975).
192. H. Wachsmuth, *J. Pharm. Belg. (N.S.)*, **5**, 300 (1950).
193. J. Holzbecher and D. E. Ryan, *Anal. Chim. Acta*, **64**, 147 (1973).
194. C. F. Brunk, K. C. Jones, and T. W. James, *Anal. Biochem.*, **92**, 497 (1979).
195. O. Dann, G. Bergen, T. Demant, and G. Volz, *Ann. Chem.*, **794**, 68 (1971).
196. C. F. Brunk and T. W. James, *J. Cell. Biol.*, **75**, 136a (1977).
197. J. W. M. Visser, A. A. M. Jongeling, and H. J. Tanke, *J. Histochem. Cytochem.*, **27**, 32 (1979).
198. D. G. Macinnes, H. W. Reading, and A. I. M. Glen, *Biochem. Soc. Trans.*, **8**, 340 (1980).
199. G. C. Cotzias and A. C. Foradori, in *The Biological Basis of Medicine*, edited by E. E. Bittar and N. J. Bittar, Academic Press, London, 1968, pp. 105–121.
200. M. D. Muir and P. R. Grant, in *Quantitative Scanning Electron Microscopy*, edited by D. B. Holt, M. D. Muir, P. R. Grant, and I. M. Boswarud, Academic Press, London, 1980, p. 287.
201. S. Nordling and S. Aho, *Anal. Biochem.*, **115**, 260 (1981).
202. H. E. Hirsch and M. E. Parks, *Anal. Biochem.*, **122**, 79 (1982).
203. H. Corrodi and B. Werdinius, *Acta Chem. Scand.*, **19**, 1854 (1965).
204. G. G. Guilbault, P. Brignac, and M. Zimmer, *Anal. Chem.*, **40**, 190 (1968).

# REFERENCES

205. O. S. Wolfbeis, E. Fürlinger, H. Kroneis, and H. Marsoner, *Fresenius Z. Anal. Chem.*, **314**, 119 (1983).
206. D. A. Priestman and J. Butterworth, *Clin. Chim. Acta*, **142**, 263 (1984).
207. C. R. Scriver, R. J. Smith, and J. M. Phang, in *The Metabolic Basis of Inherited Disease*, edited by J. B. Stanbury, J. B. Wyngaarden, D. S. Fredrickson et al., McCraw-Hill, New York, 1983, pp. 375–377.
208. F. Endo, I. Matsuda, A. Ogata, and S. Tanaka, *Paediatr. Res.*, **227** (1982).
209. P. S. Pederson, E. Christensen, and N. J. Brandt, *Acta Paediatr. Scand.*, **72**, 785 (1983).
210. J. Butterworth and D. Priestman, *J. Inher. Metab. Dis.*, **7**, 32 (1984).
211. T. Matsumoto, T. Furuta, Y. Nimura, and O. Suzuki, *Biochem. Pharmacol.*, **31**, 2207 (1982).
212. I. Hemmilä, *Clin. Chem.*, **31**, 359 (1985).
213. K. Miyai, *Bunseki*, **122**, 79 (1985).
214. E. Engvall, *Methods Enzymologia*, **70**, 419 (1980).
215. A. F. Esser, *Lab. Res. Methods Biol. Med.*, **4**, 213 (1980).
216. T. Olsson and A. Thore, in *Immunoassays for the 80's*, edited by A. Voller, A. Bartlett and D. Bidwell, MTP-Press, U.K., 1981, pp. 113–125.
217. M. Cais, S. Dani, Y. Eden et al., *Nature*, **270**, 534 (1977).
218. E. Soini, I. Hemmila, *Clin. Chem.*, **25**, 353 (1979).
219. R. M. Nakamura, *Lab. Res. Methods Biol. Med.*, **3**, 211 (1979).
220. C. M. O'Donnell and S. C. Suffin, *Anal. Chem.*, **51**, 33 (1979).
221. D. S. Smith and M. H. H. Al-Hakiem, *Ann. Clin. Biochem.*, **18**, 253 (1981).
222. J. Landon and R. S. Kamel, in ref. 216, pp. 91–112.
223. F. F. Ullman, *Tokai J. Exp. Clin. Med.*, **4**, 7 (1979).
224. A. J. Quattrone, C. M. O'Donnell, J. Mcbride et al., *J. Anal. Toxicol.*, **5**, 245 (1981).
225. G. C. Visor and S. G. Schulman, *J. Pharm. Sci.*, **70**, 469 (1981).
226. E. T. Maggio, in *Immunoassays: Clinical Laboratory Techinques for the 1980's*, edited by R. M. Nakamura, W. R. Dito, and E. S. Tucher, Alan R. Liss Inc., New York, 1980, pp. 1–12.
227. A. H. Coons, H. J. Creech, and R. N. Jones, *Proc. Soc. Exp. Biol. Med.*, **47**, 200 (1941).
228. G. I. Ekeke, D. Exley, and R. Abuknesha, *J. Steroid Biochem.*, **11**, 1597 (1979).
229. Y. Kobayashi, M. Yamata, I. Watanabe, and K. Miyai, *J. Steroid Biochem.*, **16**, 521 (1982).

230. L. A. Kaplan, N. Gau, E. A. Stein et al., *Chin. Biochem. Anal.*, **10**, 183 (1981).
231. R. E. Curry, H. Heitzman, D. H. Riege et al., *Clin. Chem.*, **25**, 1591 (1979).
232. M. Pourfarzaneh and R. D. Nargessi, *Clin. Chim. Acta*, **111**, 61 (1981).
233. Y.-G. Tsay, L. Wilson, and E. Keefe, *Clin. Chem.*, **26**, 1610 (1980).
234. R. W. Stewens, D. Elmendorf, M. Courlay et al., *J. Clin. Microbiol.*, **10**, 346 (1979).
235. R. C. Aalberse, *Clin. Chim. Acta*, **48**, 109 (1973).
236. M. J. Sinosich, *Ann. Clin. Biochem.*, **16**, 334 (1979).
237. R. P. Ekins, *Clin. Chim. Acta*, **5**, 453 (1960).
238. W. B. Dandliker, J. H. Schapiro, J. W. Meduski et al., *Immunochemistry*, **1**, 165 (1964).
239. W. B. Dandliker and S. A. Levison, *Immunochemistry*, **5**, 171 (1967).
240. R. P. Tengerdy, *J. Lab. Clin. Med.*, **70**, 707 (1967).
241. W. B. Dandliker and V. A. DeSaussure, *Immunochemistry*, **7**, 799 (1970).
242. W. B. Dandliker, R. J. Kelly, J. Dandliker et al., *Immunochemistry*, **10**, 219 (1973).
243. R. D. Spencer, F. B. Toledo, B. T. Williams, and N. L. Yoss, *Clin. Chem.*, **19**, 838 (1973).
244. C. W. Parker, T. J. Yoo, M. C. Johnson, and S. M. Godt, *Biochemistry*, **11**, 3408 (1967).
245. R. P. Liburdy, *J. Immunol. Methods*, **28**, 233 (1979); and U.S. Patent 4207075 (1980).
246. G. Handley, J. N. Miller, and J. W. Bridges, *Proc. Anal. Div. Chem. Soc.*, **16**, 26 (1979).
247. R. Rezi-Poor-Kardost and E. W. Voss, *Mol. Immunol.*, **19**, 159 (1981).
248. R. M. Watt and E. W. Voss, *Immunochemistry*, **14**, 533 (1977).
249. E. J. Shaw, R. A. A. Watson, and D. S. Smith, *Clin. Chem.*, **25**, 322 (1979).
250. E. F. Ullman, M. Schwarzberg, and K. E. Rubenstein, *J. Biol. Chem.*, **251**, 4172 (1976).
251. J. N. Miller, C. S. Lim, and J. N. Bridges, *Analyst*, **105**, 91 (1980).
252. H. Thakrar and J. N. Miller, *Anal. Proc.*, **19**, 329 (1982).
253. P. L. Khanna and E. F. Ullman, *Anal. Biochem.*, **108**, 15 (1980).
254. P. C. Khanna and E. F. Ullman, Eur. Patent 50684 (1982).
255. D. J. Litman, Z. Harel, and E. F. Ullman, U.S. Patent 4318707 (1982).
256. G. Steinbach and H. Mayersbach, *Acta Histochem.*, **55**, 110 (1976).
257. K. Watzek, German Patent 2537275 (1977).
258. J. F. Burd, R. J. Carrilo, M. C. Fetter et al., *Anal. Biochem.*, **77**, 56 (1977).
259. F. Kohen, Z. Hollander, J. F. Burd, and R. C. Boguslaski, *FEBS Lett.*, **100**, 137 (1979).
260. M. N. Kronick and W. A. Little, *J. Immunol. Methods*, **8**, 235 (1975).

261. P. J. Lisi, C. W. Huang, R. A. Hoffman, and J. W. Tempel, *Clin. Chim. Acta*, **120**, 171 (1982).
262. H. M. McConnell, German Patent 2938646 (1980).
263. J. Briggs, V. B. Elings, and D. F. Nicoli, *Science*, **212**, 1266 (1981).
264. S. D. Lidofsky, T. Imasaka, and R. N. Zare, *Anal. Chem.*, **51**, 1602 (1979).
265. A. B. Bradley and R. N. Zare, *J. Am. Chem. Soc.*, **98**, 620 (1976).
266. T. Imasaka, H. Kadone, T. Ogawa, and N. Ishibashi, *Anal. Chem.*, **49**, 667 (1977).
267. B. W. Smith, F. W. Plankey, N. Omenetto et al. *Spectrochim. Acta*, Part A **30**, 1459 (1974).
268. J. H. Richardson, B. W. Wallin, D. C. Johnson, and L. W. Hrubesh, *Anal. Chim. Acta*, **86**, 263 (1976).
269. M. Goldman, *Fluorescence Antibody Methods*, Academic Press, New York, 1968, pp. 97–99.
270. D. Blakeslee and M. Baines, *J. Immunol. Methods*, **13**, 305 (1976).
271. M. Goldman, *Fluorescence Antibody Methods*, Academic Press, New York, 1968, pp. 129–130.
272. Y. Ikariyama, S. Suzuki, and M. Aizawa, *Anal. Chem.*, **54**, 1126 (1982).
273. J. S. Woodhead, I. Weeks, A. K. Campbell et al., in *Luminescent Assays: Perspectives in Endocrinology and Clinical Chemistry*, edited by M. Serio and M. Pazzagli, Raven Press, New York, 1982, pp. 147–155.
274. I. Weeks, I. Beheshti, F. McCapra et al., *Clin. Chem.*, **29**, 1474 (1983).
275. I. Weeks, A. K. Campbell and J. S. Woodhead, *Clin. Chem.*, **29**, 1480 (1983).
276. A. Patel and A. K. Campbell, *Clin. Chem.*, **29**, 1604 (1983).
277. T. P. Whitehead, G. H. G. Thorpe, T. J. N. Carter et al., *Nature*, **305**, 158 (1983).
278. W. Klingler, C. J. Strasburger, and W. G. Wood, *Trends Anal. Chem.*, **2**, 132 (1983).
279. H. R. Schroeder, R. C. Boguslaski, R. J. Carrico, and R. T. Buckler, in *Methods in Enzymology*, edited by M. DeLuca and W. McElroy, Vol. LVII, Academic Press, New York, 1978, p. 424.
280. F. Kohen, J. B. Kim, H. R. Lindner, and W. P. Collins, *Steroids*, **38**, 73 (1981).
281. M. Pazzagli, J. B. Kim, G. Messeri et al., *Clin. Chim. Acta*, **115**, 277 (1981).
282. M. Pazzagli, J. B. Kim, G. Messeri et al., *Clin. Chim. Acta*, **115**, 287 (1981).
283. C. J. Strasburger, H. Feicke, and W. G. Wood, *Fresenius Z. Anal. Chem.*, **311**, 351 (1982).
284. M. Pazzagli, J. B. Kim, G. Messeri et al., *J. Steroid Biochem.*, **14**, 1005 (1981).

285. M. Pazzagli, J. B. Kim, G. Messeri et al., *J. Steroid Biochem.*, **14**, 1181 (1981).
286. F. Kohen, M. Pazzagli, J. B. Kim, and H. R. Lindner, *Steroids*, **36**, 421 (1980).
287. H. R. Schroeder, F. M. Yeager, R. C. Boguslaski, and P. O. Vogelhut, *J. Immunol. Methods*, **25**, 275 (1979).
288. Z. Eshhar, J. B. Kim, G. Barnard et al., *Steroids*, **38**, 89 (1981).
289. F. Kohen, J. B. Kim, G. Barnard, and H. R. Lindner, *Steroids*, **36**, 405 (1980).
290. M. Pazzagli, M. Serio, P. Munson, and D. Rodbard, in *Proceedings of the International Symposium on Radioimmunoassay and Related Procedures in Medicine*, IAEA-SM-259/13, Vienna, 1982.
291. W. Klingler, G. von Postel, O. Haupt, and R. Knuppen, *Fresenius Z. Anal. Chem.*, **311**, 352 (1982).
292. W. Klingler, O. Haupt, G. von Postel, and R. Knuppen, *Acta Endocrinol. (Copenhagen) Suppl.*, **246**, 22 (1982).
293. H. Shah, A.-M. Saranko, M. Hörkönen, and H. Adlercreutz, *Clin. Chem.*, **30**, 185 (1984).
294. A. Tommasi, M. Pazzagli, M. Damiani et al., *Clin. Chem.*, **30**, 1597 (1984).
295. J. D. Boever, F. Kohen, D. Vandekerckhove, and G. V. Maele, *Clin. Chem.*, **30**, 1637 (1984).
296. S. K. Chakrabarti, J. Brodeur, and R. Tardif, *Clin. Chem.*, **21**, 1783 (1975).
297. D. Bonsignore, P. Calissano, and C. Cortasangna, *Med. Lav.*, **56**, 199 (1965).
298. H. B. Burch and A. L. Siegel, *Clin. Chem.*, **17**, 1038 (1971).
299. J. B. Weissberg, F. Lipschutz, and F. A. Oski, *New Engl. J. Med.*, **284**, 565 (1971).
300. K. D. Gibson, A. Neuberger, and J. J. Scott, *Biochem. J.*, **61**, 618 (1955).
301. S. Granick and D. Mauzerall, *J. Biol. Chem.*, **232**, 1119 (1958).
302. E. J. Majeski, E. J. Seltzer, P. L. Carter et al., *Clin. Chem.*, **23**, 1976 (1977).
303. F. R. Elevitch, *Fluorometric Techniques in Clinical Chemistry*, Little, Brown and Co., Boston, Mass, 1973, p. 250.
304. E. Salkowski, *Arch. Ges. Physiol.*, **6**, 207 (1972).
305. S. Krastelewsky, *Biochemistry*, **143**, 403 (1923).
306. D. Abramsohn, *Biochem. Z.*, **198**, 233 (1928).
307. V. O. Merkelbach, *Helv. Med. Acta*, **10**, 67 (1943).
308. C. Dhéré and L. Laszt, *C.R. Soc. Biol.*, **143**, 87 (1949).

309. L. L. Abell and B. B. Levy, *Standard Methods Clin. Chem.*, **2**, 26 (1958).
310. L. L. Abell, B. B. Levy, B. B. Brodie, and F. E. Kendall, *J. Biol. Chem.*, **195**, 357 (1952).
311. J. de la Huerga and J. C. Sherrick, *Ann. Clin. Lab. Sci.*, **2**, 360 (1972).
312. S. K. Chakrabarti, *Clin. Chem.*, **25**, 592 (1979).
313. H. Ohtsuji and M. Ikeda, *Br. J. Ind. Med.*, **27**, 150 (1970).
314. J. P. Buchet, R. Lauwerys, and H. Roels, *Arch. Mal. Prof. Med. Trav. Secur. Soc.*, **25**, 511 (1974).
315. K. Engström and J. Rantanen, *Int. Arch. Arbeitsmad.*, **33**, 163 (1974).
316. D. Bauer and M. Guillemin, *Int. Arch. Occup. Environ. Health*, **37**, 47 (1976).
317. M. Guillemin and D. Bauer, *Int. Arch. Occup. Environ. Health*, **37**, 57 (1976).
318. K. H. Schaller, H. W. Schütz, V. Geldmacher et al., *Acta Pharmacol. Toxicol.*, **41**, 230 (1977).
319. N. Filipescu, S. K. Chakrabarti, and P. G. Tarassoff, *J. Phys. Chem.*, **77**, 2276 (1973).
320. A. A. Lamola, *J. Invest. Dermatol.*, **77**, 114 (1981).
321. W. C. Maslow, H. A. Muensch, F. Azama, and A. S. Schneider, *Clin. Chem.*, **29**, 260 (1983).
322. J. Bramhall and A. Z. Britten, *Clin. Chim. Acta*, **68**, 203 (1976).
323. G. H. Beastall and S. McVeigh, *Clin. Chim. Acta*, **70**, 343 (1976).
324. J. B. Brown, S. C. Macleod, C. Macnaughton et al., *J. Endocrinol.*, **42**, 5 (1968).
325. J. B. Brown and N. A. Beischer, *Obstet. Gynecol. Survey*, **27**, 205 (1972).
326. M. Lebeau and J. Van Peborgh, *J. Gynecol. Obstet. Biol. Reprod.*, **2**, 63 (1973).
327. R. F. Straw and F. W. Hanson, *J. Med.*, **6**, 369 (1975).
328. R. Frye, P. F. Fong, G. F. Johnson, and R. C. Rock, *Clin. Chem.*, **23**, 1819 (1977).
329. R. Hahnel, and H. Jones, *Clin. Chim. Acta*, **16**, 185 (1967).
330. J. E. Hammond, J. C. Phillips, and J. Savory, *Clin. Chem.*, **24**, 631 (1978).
331. B. E. Dalrymple, J. M. Duff, and E. R. Menzel, *J. Forensic Sci.*, **22**, 106 (1977).
332. R. D. Olsen, *Fingerprint and Identification Magazine*, "The Oils of Latent Fingerprints," Vol. 56, No. 7, pp. 3–12 (1975).
333. R. D. Olsen, *Fingerprint and Identification Magazine*, "The Chemical Composition of Palmar Sweat," Vol. 53, No. 10, pp. 3–23 (1972).
334. G. G. Guilbault, *Practical Fluorescence—Theory, Methods, and Techniques*, M. Dekker, New York, 1973, p. 334.

335. M. Stöhr and M. Vogt-Schaden, *Acta Pathol. Microbiol. Scand., Suppl.*, **274**, 96 (1981).
336. K.-P. Völkl, N. Opitz, and D. W. Lübbers, *Fresenius Z. Anal. Chem.*, **301**, 162 (1980).
337. N. Lakshminarayanaiah, *Membrane Electrodes*, Academic Press, New York, 1976.
338. W. M. Vaughan and G. Weber, *Biochemistry*, **9**, 464 (1970).
339. H. Sham and F. Wagner, *Eur. J. Biochem.*, **36**, 250 (1973).

CHAPTER

5

# BIOCHEMICAL AND BIOMEDICAL APPLICATIONS OF FLUOROMETRIC ANALYSIS USING HPLC

N. ICHINOSE, F.-M. SCHNEPEL, G. SCHWEDT, AND K. ADACHI

## 1. BIOCHEMISTRY

### 1.1. Amines, Amino Acids, and Related Compounds

#### 1.1.1. Amines (General)

The stepwise fluorometric determination of primary and secondary amines based on precolumn derivatization and high-performance liquid chromatography (HPLC) separation is described by Nakamura et al. (1). The amines are reacted with 2-methoxy-2,4-diphenyl-3(2$H$)-furanone (MDPF; see Chapter 4, Section 1.1) at pH 9.6 and 20°C for 30 min to produce fluorescent pyrrolinones from primary amines and nonfluorescent aminodienones from secondary amines. The MDPF derivatives are separated on a reversed-phase $C_{18}$ column with a mixture of methanol and phosphate buffer (70 : 30) (pH 7.0). After detection of the pyrrolinones with the first fluorescence monitor ($\lambda_{ex} = 360$ nm, $\lambda_{em} > 405$ nm), the eluate is mixed with 12 mol dm$^{-3}$ ethanolamine hydrochloride (pH 10.5) to convert the aminodienones to fluorescent MDPF-ethanolamine, which is detected with the second fluorescence monitor ($\lambda_{ex} = 390$ nm, $\lambda_{em} = 480$ nm) (for reaction pathways see Fig. 4.3$b$). The method permits the determination of 3 pmol lower $n$-alkylamines and 50 pmol lower di-$n$-alkylamines.

Another application of precolumn derivatization with 7-fluoro-4-nitrobenzo-2-oxa-1,3-diazole (NBD-F; see Fig. 4.2$e$) is reported by Toyooka et al. (2). Amines, amino acids, imino acids, and polyamines are derivatized and separated on columns of polyethyleneglycol dimethacrylate gel (for further details see Section 1.1.3). The reaction rates decrease in the following order; secondary amines, imino acids, primary amines, amino acids. All amino compounds tested, except tryptophan, gave fluorescent derivatives with excitation maxima of 467–472 nm and emission maxima at 524–541 nm.

The reaction of amines with $o$-phthalaldehyde (see Chapter 4, Section 1.1) was utilized in several HPLC applications. Two of them will be discussed in more detail. The basic disadvantage with postcolumn derivatization systems is the introduction of additional band broadening, which occurs in the flow-through reactor. This problem was limited by using a zigzag open-tubular capillary, where plate height decreased with the square of the capillary diameter (3). The reactor was used for a model fluorescence derivatization reaction, a modification of the very fast *o*-phthalaldehyde reaction with primary amines, which incorporates the use of 3-mercaptopropionic acid. The system was applied to the analysis of amino acids and primary aliphatic amines after separation on an octadecyl-silica reversed-phase column.

Primary and secondary amines were determined simultaneously with a HPLC method described by Himuro et al. (4). The amines are separated on packed columns of a strong cation exchanger bonded to silica gel, and the postcolumn derivatization is based on the manual procedure of Himuro et al., described in Chapter 4, Section 1.1. Secondary amines are converted to primary amines with NaClO, and then primary amines are derivatized with *o*-phthalaldehyde-2-mercaptoethanol reagent with excess NaClO suppressed by 2,2'-thiodiethanol. This conversion-derivatization-fluorometric detection method was studied for typical secondary amines and analogous primary amines by flow-injection analysis.

Another example of postcolumn derivatization of amines is the ion-pair reaction reported by Brinkman et al. (5). After their HPLC separation, tertiary amines are converted to fluorescent ion pairs with the reagent 9,10-dimethoxyanthracene-2-sulfonate. A comparison of the air-segmentation and solvent-segmentation approaches shows no significant difference regarding band broadening. The technique is feasible for normal- and reversed-phase HPLC, and can be applied, for example, to the determination of chlorpheniramine in human urine samples.

### *1.1.2. N-Heterocyclic Compounds*

As described in Chapter 4, Section 1.1, certain aromatic and heterocyclic amines show native fluorescence that is sufficiently intense for their determination. Therefore, native fluorescence can also be utilized for the detection of certain amines after HPLC separation. The simultaneous determination of indoles in the tryptamine pathway (tryptamine, indoleacetic acid, tryptophan, tryptophol) in mouse-brain samples is described by Yamada et al. (6). The HPLC system consists of a $C_8$ reversed-phase column, a mobile phase of acetate buffer (pH 5.0) containing 35% methanol, and a fluorometer with excitation and emission wavelengths of 280 nm and 350 nm, respectively. Detection limits range from 10 to 20 pg, and recoveries are 86–90%.

Assenza and Brown (7) report a method to enhance the native fluorescence of purines and pyrimidines by using a postcolumn reactor. Purines and pyrimidines undergo changes in their absorption spectra depending on pH, and under certain conditions emit appreciable fluorescence after UV excitation, the fluorescence being attributed to the ionized molecules formed in very acidic or basic media. The HPLC separation was performed on reversed-phase columns of the octadecylsilica type. Eluants for isocratic elution were phosphate buffers (pH 3.5–6.5), and for gradient elution, solutions of $KH_2PO_4$ containing 0–24% methanol were used. As postcolumn reagents, solutions of two acids (sulfuric acid and phosphoric acid) and two bases (potassium hydroxide and sodium hydroxide) were evaluated. The excitation wavelength was scanned from 220 to 300 nm, while the emission was monitored at 370 nm. With fluorescence detection, some methylated purines were determined at levels as low as 1 pmol. The method has advantages: the protonation and deprotonation of the molecules is rapid; and as the method is nondestructive, chromatographic fractions can be collected for further analysis.

The determination of free and conjugated pteridines in human blood (plasma, erythrocytes, lymphocytes, buffy coat) by reversed-phase HPLC is described by Zeitler et al. (8). Ten pteridines (pterin-6-carboxylic acid, xanthopterin, neopterin, monapterin, isoxanthopterin, lumazine, biopterin, 6-hydroxymethylpterin, pterin, and 6-methylpterin) are separated on a $C_{18}$ column with elution by a $KH_2PO_4$ buffer (pH 7.1–7.8) and determined with fluorometric detection in the femtomole range.

The following methods for the determination of certain aromatic amines are based on their derivatization with *o*-phthalaldehyde or other fluorogenic reagents (see also Chapter 4, Section 1.1). As discussed in Chapter 4, Sections 1.1 and 1.2, amines and amino acids are derivatized to fluorescent products by the reagent *o*-phthalaldehyde in combination with a thiol compound like mercaptoethanol. Several amines, however, notably histamine, do not require the thiol; by reaction alone with *o*-phthalaldehyde in alkaline media, a product is formed that gives a highly increased fluorescence upon acidification. Using this reaction as a postcolumn derivatization after HPLC separation, Allenmark et al. (9) developed a selective method for the determination of histamine in biological samples. The separation is performed on a $C_{18}$ reversed-phase material with a mobile phase of 5 mmol dm$^{-3}$ pentane sulfonic acid in citrate buffer, containing 5% methanol. The method allows the determination of histamine even at low-picomolar levels.

The derivatization with $o$-phthalaldehyde was also applied to a continuous-flow method, by which histamine can be determined in blood plasma and also in a mixture of polyamines, such as putrescine, spermidine, and spermine (10). Before to the HPLC separation, the plasma samples (200 mm$^3$) are deproteinized with $ClO_4^-$ and extracted with buthanol and $n$-heptane. HPLC is performed on a column packed with Hitachi gel 3011-C, and as mobile phase 0.27 mol dm$^{-3}$ $KH_2PO_4$ and 0.35 mol dm$^{-3}$ $KH_2PO_4$ solutions are used. Histamine can be separated from a mixture of polyamines with detection limits of about 25 ng cm$^{-3}$.

A new selective method for quantitation of adenosine in canine myocardial extracts is described by Jacobson et al. (11). The method involves incubation of the extracts with chloroacetaldehyde to form the fluorescing adenosine derivative $1,N^6$-ethenoadenosine. The ethenoadenosine was separated by reversed-phase HPLC on Ultrasphere-ODS and quantitated by fluorometry. The results for myocardial adenosine contents with the fluorometric method were nearly identical to those obtained by the routine technique of HPLC with direct detection by UV absorption. A primary advantage of the fluorometric method, however, is its greater sensitivity: as little as 0.50 pmol on the column could be quantitated by fluorescence, compared with 20 pmol with UV absorption.

Based on this derivatization, a sensitive and specific assay for measurement of adenine nucleotides and adenosine by ion-pair HPLC is described by Ramos-Salazar and Baines (12). The $1,N^6$-etheno derivatives of ATP, ADP, AMP, and adenosine, formed by reaction with chloroacetaldehyde, were separated under isocratic conditions in 20 min. These compounds are strongly fluorescent at an emission wavelength of 280 nm, rendering a lowest detection limit of 2–5 pmol per injection. Specificity was confirmed enzymatically. The method was applied to determine adenine nucleotides and adenosine in oxygenated and hypoxic perfused rat kidneys.

A sensitive method to determine adenosine compounds simultaneously was established by introducing a new fluorescent reagent into HPLC (13). Bromoacetaldehyde, the most suitable reagent among the haloacetaldehydes examined, shows quantitative reaction even with unstable ADP and ATP. A high resolution of adenine nucleotide was obtained with a gel column and elution with a solvent consisting of citric acid–$Na_2HPO_4$–NaCl buffer (pH 5.0) and methanol (1 : 1). The method was applied to the measurement of cAMP in urine, and of ADP and ATP in brain and blood. Femtomole amounts of the adenine nucleotides were clearly separated.

### 1.1.3. Polyamines

Polyamines, too, are usually derivatized with o-phthalaldehyde/thiol or with one of the other reagents discussed in Chapter 4, Section 1.1. Some recent applications are described in more detail. The separation of natural polyamines and their monoacetyl derivatives by reversed-phase HPLC is reported by Seiler and Knoedgen (14). To form ion pairs with the polycations, octane sulfonate was used, and the o-phthalaldehyde method was applied for postcolumn derivatization. The method allows the determination of polyamine and acetylspermidine directly from tissue extracts and body fluids without prepurification.

Simpson et al. (15) report a similar ion-pair HPLC procedure for putrescine, spermidine, and spermine in urine and serum samples. As ion-pair reagent, heptanesulfonate is employed, and o-phthalaldehyde/2-mercaptoethanol used for on-line postcolumn derivatization and subsequent fluorescence detection. The detection limits for the polyamines range from 120 pmol for spermine to 12 pmol for putrescine. The method includes a gradient program that provides complete separation from amino acids and specificity for the three polyamines. The effects of several variables, such as pH, concentration of the aqueous buffer, counter-ion concentration, and percentage of organic modifier in the moving phase, are discussed.

For precolumn derivatization of polyamines, other reagents have also been applied. A rapid and simple method for the determination of putrescine, spermidine, and spermine in biological samples is described by Minchin and Hanau (16). The procedure involves protein precipitation with $HClO_4$ and dansylation of the polyamines. After extraction on a $C_{18}$ cartridge, the samples are analyzed by HPLC, using a step solvent change and a $C_{18}$ reversed-phase column. The chromatographic conditions allow complete analysis of the three polyamines within 10 min. Standard curves are linear up to 1 µg polyamine, and the relative standard deviation for the assay ranged from 4% at 1 µg polyamine per sample to 11% at 50 ng polyamine per sample. The reported technique, which does not require dual pumps, ion-pairing agents, solvent extractions, or a gradient control system, was applied to the determination of putrescine, spermidine, and spermine in rat lung, liver, and kidney.

Polyamines with primary or secondary amino groups in the skeleton (e.g., putrescine, spermidine, and spermine) are derivatized rather quickly with 4-fluoro-7-nitrobenzo-2-oxa-1,3-diazole (NBD-F; see Fig. 4.2e) to the $N,N$-di-, $N,N,N$-tri-, and $N,N,N,N$-tetra-NBD derivatives, respectively (2). The HPLC separation of the three derivatives is performed on a polyethyleneglycol dimethylacrylate gel column, using methanol/acetonitrile, methanol/acetonitrile/HCl, and methanol/phosphate buffer (pH 6) as eluants. The detection limits are 76 fmol (putrescine), 0.43 pmol (spermidine), and 1.20 pmol (spermine).

$\alpha,\omega$-Diamines in putrefied meat samples were determined after precolumn derivatization of fluorescent bis(diphenylboron)chelates (see Section 3).

### 1.1.4. Amino Sugars

Two methods for the analysis of hexosamines by HPLC and postcolumn derivatization are presented by Honda et al. (17, 18).

For the first application, glucosamine and galactosamine are separated on a polystyrene column, sulfonate type, with a borate buffer (pH 7.5) containing NaCl. The hexosamines in the eluate are monitored fluorometrically at 331 (excitation) and 383 (emission) nm, respectively, postcolumn labeling with 2-cyanoacetamide, and also photometrically (276 nm). This simple method allows the simultaneous, automated determination of 10–500 nmol glucosamine and galactosamine with high reproducibility. The method was applied to the determination of hexosamines in some commercial glycoconjugates and in human urine.

The second method for the analysis of glucosamine and galactosamine involves postcolumn labeling with 6% 2,4-pentanedione and 9% formaldehyde in $0.10 \, \text{mol dm}^{-3}$ acetate buffer (pH 4.8) and fluorometric detection, following hydrolysis of the glycoconjugates (6 h in $4 \, \text{mol dm}^{-3}$ HCl at 100°C under a nitrogen atmosphere). Under the conditions used, 20–500 pmol hexosamines could be determined with high reproducibility. The lower detection limits for glucosamine and galactosamine were 140 and 230 pmol, respectively, and the method was more sensitive by approximately one order of magnitude than that involving labeling with 2-cyanoacetamide. Amino acids did not interfere with the determination of hexosamine.

The microquantitative analysis of neutral and amino sugars in glycoconjugates was developed by Takemoto et al. (19). The glycoconjugates were hydrolyzed with a mixture of trifluoroacetic acid and HCl, and the free amino groups were acetylated. Sugars were coupled with 2-aminopyridine. The excess reagents were removed by high-performance gel chromatography, and the fluorescent pyridylamino derivatives of the sugars were separated and quantified by reversed-phase HPLC. This method allows the determination of 0.01–10 nmol sugar.

### 1.1.5. Amino Acids and Imino Acids

A general review of postcolumn derivatizations of amino acids and polypeptides after HPLC separation is presented by Lewis (20).

Six traditional methods for the HPLC analysis of amino acids are compared by Dong et al. (21). These methods include the following:

1. Ion exchange postcolumn derivatization with:
   a. ninhydrin;
   b. *o*-phthalaldehyde/hypochlorite (see Chapter 4, Section 1.1).
2. Reversed-phase HPLC and precolumn derivatization with:
   a. dansyl chloride (see Fig. 4.2*d*);
   b. *o*-phthalaldehyde (see Fig. 4.2*b*, and Section 1.1 of this chapter),
   c. phenylthiohydantoin.

For precise quantitation, postcolumn derivatization is best; use of *o*-phthalaldehyde resulted in higher sensitivity than was achieved with ninhydrin. Because of its high resolution, short analysis time, and high sensitivity, precolumn derivatization with *o*-phthalaldehyde was suggested for food analysis and clinical chemistry.

### 1.1.6. Some Important Applications

**1.1.6.1. Derivatization with *o*-Phthalaldehyde.** As described in Chapter 4, Section 1.2, *o*-phthalaldehyde reacts rapidly with primary amino acids in the presence of a thiol compound to form intensely fluorescent derivatives. In combination with HPLC separations, this derivatization can be used either as precolumn or as postcolumn reaction (see Table 5.1). During the last years, several applications based on this method have been published, some of which are discussed in this chapter.

*1.1.6.1.1. Precolumn Derivatization.* A fluorometric procedure for the quantitative amino acid analysis of a typical peptide hydrolysate is presented by Larsen and West (22). The amino acids are derivatized

Table 5.1. Some HPLC Methods for the Analysis of Amino and Imino Acids Using the Derivatization Reaction with o-Phthalaldehyde/thiol Reagent

| Compound | Column Material | Eluant | Detection Limit | References |
|---|---|---|---|---|
| *a. Precolumn Derivatization* | | | | |
| Amino acids in peptide hydrolyzates | $C_{18}$/fatty acid analysis column | Triethylamine acetate/acetonitrile | ~25 pmol | 22 |
| Amino acids in blood plasma, muscle, liver | RP-Ultrasphere ODS | Sodium acetate/methanol | <100 fmol | 23 |
| | RP-Supersphere CH-8 Spherisorb ODS II | Sodium phosphate/acetonitrile | 1–10 pmol | 24 |
| Amino acids in human plasma, cerebrospinal fluids, gastric juice, rat brain | RP-Hypersil ODS | Methanol/sodium phosphate | | 25 |
| Amino acids in blood plasma | RP-18 Lichrosorb | Methanol-sodium acetate/methanol-sodium citrate | | 26 |
| Imino acids | RP-Ultrasphere ODS | Phosphate-propionate-acetonitrile-water/acetonitrile-methanol water | | 27 |
| *b. Postcolumn Derivatization* | | | | |
| Amino acids, imino acids | Cation exchange | Sodium citrate, lithium citrate | 5 pmol (primary amines), 100 pmol (secondary amines) | 28 |

with o-phthalaldehyde before the HPLC separation, and the adducts are resolved using a linear gradient of acetonitrile against an initial solvent consisting of a triethylamine acetate solution (pH 7.5). Complete analysis requires 46 min, and the lower limit of quantitation is 100 pmol, with a lower detection limit of 25 pmol.

Another example of the analysis of peptide hydrolysates (and also physiological fluids) was published by Jones and Gilligan (23). After the precolumn derivatization with o-phthalaldehyde, the derivatives are analyzed with good selectivity by HPLC, employing 3-$\mu$m particle size, reversed-phase columns. The amino acid derivatives are resolved with a methanol gradient in $0.1 \, \text{mol dm}^{-3}$ aqueous sodium acetate (pH 7.2). The quantitations of the individual amino acid derivatives are reproducible within an average relative deviation of $\pm 1.5\%$ and have a detection limit of $< 100$ fmol. Amino acid mixtures, which are obtained by enzymic or acid hydrolysis of polypeptides, are efficiently resolved within less than 18 min. Methods for the amino acid analysis of physiological fluids such as serum, urine, and cerebrospinal fluid are also presented. Mixtures that contain as many as 48 components can be resolved within less than 50 min.

The precolumn derivatization and HPLC separation can be applied also to the analysis of serum samples. Fleury and Ashley (29) present a method for the determination of 21 amino acids in blood serum. This method includes an automatic on-line precolumn procedure for derivatization, permits full automation, and avoids problems due to time-dependent decay of fluorescence of the derivatives. The total time for analysis is less than 60 min, and limits of sensitivity are approximately 100 fmol. Proline, hydroxyproline, and cysteine were not detected. Comparison with results obtained with a standard amino acid analyzer shows the reliability of this method.

Elkin (30) describes the derivatization of free amino acids in chick, turkey, and duck serum samples by HPLC and compares this method to ion-exchange chromatography. The HPLC separation is performed with a reversed-phase system ($\mu$ Bondapak $C_{18}$) after precolumn derivatization with o-phthalaldehyde-ethanethiol. By this method 17 amino acids are separated and quantitated within 70 min (including column regeneration). By comparison, the same serum samples were analyzed by conventional cation-exchange chromatography on an amino acid analyzer, using postcolumn derivatization with ninhydrin and absorbance measurement. This method requires about 270 min. Within a species, the serum amino acid concentrations obtained by these two methods are very similar with the following exceptions: the levels of asparagine and lysine were consistently higher when determined by cation-exchange chromatography, whereas tryptophan values were consistently higher when determined by HPLC. Thus, the determination of free amino acids in avian serum can be both accurately and reproducibly achieved by precolumn derivatization and HPLC separation.

The use of HPLC for the rapid and precise determination of amino acids in human plasma is reported by Hogan et al. (31). Human plasma, deproteinized and then derivatized by $o$-phthalaldehyde, produces individual amino acid peaks in less than 35 min. The quantitative analysis of microamounts of free amino acids gives highly reproducible results. There is also good agreement between measurement of plasma amino acids with HPLC and measurement with amino acid analyzers.

Free amino acids in blood plasma were determined by Liu et al. (26), using reversed-phase HPLC with gradient elution and fluorescence detection. The blood plasma samples were deproteinized with ethanol and centrifuged; supernatants were incubated with $o$-phthalaldehyde and 2-mercaptoethanol at 25°C for 90 s, separated on LiChrosorb RP-18 column, and detected fluorometrically. Linear gradient elution was performed with two solvent systems (solvent A: methanol—0.01 mol dm$^{-3}$ sodium acetate, pH 7.0; solvent B: methanol—0.0375 mol dm$^{-3}$ sodium citrate, pH 7.0; 0–70% of solvent A was reached in 40 min). Twenty-two amino acids were determined in one run in 50 min, with recoveries of about 100%.

The determination of amino acids in blood plasma, muscle, and liver samples was also studied by Godel et al. (24). By precolumn derivatization with $o$-phthalaldehyde/3-mercaptopropionic acid, reversed-phase HPLC on Superspher CH-8 or Spherisorb ODS II, and gradient elution with methanol/sodium phosphate buffer, excellent separation of 26 physiological amino acids was obtained in the range of 1–10 pmol. The reported excitation wavelength of the derivatives is 334.5 ± 1.5 nm (S.D.), range 332–338 nm, with the exception of lysine (311 nm); the mean emission wavelength is 445 ± 2.06 nm (S.D.), range 440–449 nm. The reproducibility of the retention times is better than 1.0% for most amino acids, and the time required for analysis is less than 40 min. The results from this method are very similar to those of conventional amino acid analysis.

Amino acids in human plasma, cerebrospinal fluid, and gastric juice, and in rat-brain regions, were detected by reversed-phase HPLC and fluorometric and electrochemical detection in series (25). The $o$-phthalaldehyde-mercaptoethanol reagent was used for precolumn derivatization. HPLC was carried out on a Hypersil ODS column with isocratic or step-gradient elution, with 0.1 mol dm$^{-3}$ sodium phosphate buffer (pH 7.0)-methanol (50 : 50) or different proportions of sodium phosphate buffer (pH 5.5) and methanol, respectively. With isocratic elution, good resolution of the more hydrophobic amino acid derivatives was obtained within 20 min. For resolution of the full range of amino acids, gradient elution was used. The relation of electrochemical and fluorescent activity to increasing peptide length was also examined.

Precolumn derivatization with o-phthalaldehyde can also be used for the rapid determination of the imino acids hydroxyproline and proline (27). L-Hydroxyproline and L-proline solutions are incubated with chloramine T, followed by sodium borohydride, and then mixed with o-phthalaldehyde-2-mercaptoethanol reagent (see Fig. 5.1) and separated by reversed-phase HPLC on 5 μm Ultrasphere ODS with fluorometric detection. Gradient elution is applied with phosphate/propionate-acetonitrile-water (40 : 8 : 52) and acetonitrile-methanol-water (30 : 30 : 40). The amino acid derivatives show fluorescence intensities similar in magnitude to those of amino acid derivatives.

*1.1.6.1.2. Postcolumn Derivatization.* Postcolumn derivatization of amino acids with o-phthalaldehyde was compared to that with ninhydrin (28). By means of conventional HPLC, the amino acids were separated on strong cation exchangers and then derivatized. Detection limits are about 5 pmol for primary amines and 100 pmol for secondary amines when o-phthalaldehyde is used, and about 100 pmol for both groups with ninhydrin.

For the detection of low-picomole amounts of amino acids, including proline, a HPLC method is described by Kondo et al. (32). The procedure is based on o-phthalaldehyde postlabeling with a nonswitching NaClO flow system. With a three-solvent gradient mixer, 17 amino acids are determined in 85 min. The optimum conditions of the NaClO and o-phthalaldehyde solutions are 0.002% (0.22 cm$^3$ min$^{-1}$)

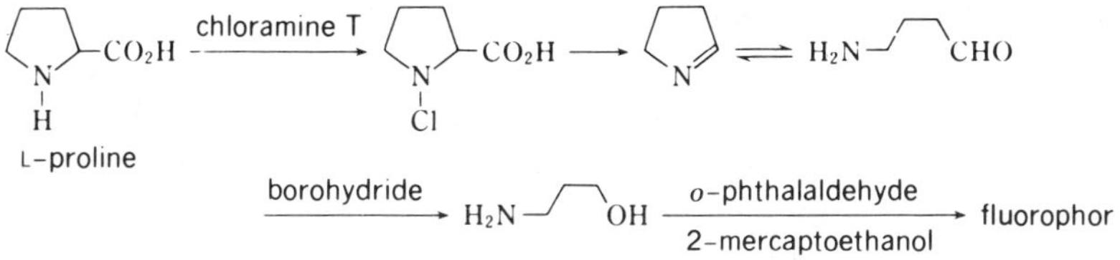

**Figure 5.1.** Reaction pathway for the derivatization of L-proline.

and 0.16% (0.26 cm$^3$ min$^{-1}$), respectively. The use of NaClO causes only 30% decrease in the fluorescence response of the usual amino acid extract for proline and cysteine, the latter being enhanced about 10-fold. The detection limits for proline and cysteine are 500 and 1000 fmol, respectively, and the limit for the usual amino acids is about 100 fmol. The calibration curves of all amino acids show good linearity in the range 5–500 pmol.

The postcolumn derivatization with $o$-phthalaldehyde and its modification, based on the conversion of secondary to primary amino groups (see Chapter 4, Section 1.1, and Section 1.1.1 of this chapter), were applied by Himuro et al. (4) to the determination of amino acids, including L-proline and L-4-hydroxyproline, as well as $N$-methyl amino acids, catecholamine, and their 3-$O$-methyl derivatives.

**1.1.6.2. Derivatization with Other Reagents.** In addition to $o$-phthalaldehyde, some of the fluorogenic reagents discussed in Chapter 4, Sections 1.1 and 1.2, have also been applied for precolumn and postcolumn derivatization in combination with HPLC techniques.

The use of the reagent $N,N$-dimethylamino-naphthalene-1-sulfonyl chloride (dansyl chloride) for the determination of amino acids by HPLC is described by Deyl and Rosmus (33), Engelhardt et al. (34), and Bayer et al. (35).

Derivatization with fluorescamine, as described in Chapter 4, Section 1.1, can cause problems for the evaluation, because two reaction products for each amino acid have been observed (36).

The reagent 7-fluoro-4-nitrobenzo-2-oxa-1,3-diazole (NBD-fluoride) was used as a precolumn fluorescent labeling reagent for HPLC of amino acids, including proline and hydroxyproline (37). The derivatization reaction is run at pH 8.0 at 60°C for 5 min. The fluorophors (aspartate, glutamate, hydroxyproline, serine, glycine, threonine, alanine, proline) are separated on a reversed-phase column ($\mu$ Bondapak C$_{18}$) with 0.1 mol dm$^{-3}$ phosphate buffer (pH 6.0) containing 6.75% methanol and 1.8% tetrahydrofuran. Detection limits are about 10 fmol with excitation at 470 nm and emission at 530 nm.

This method, based on precolumn derivatization, was applied to determine the amino acids in a few micrograms of protein hydrolysates, rabbit pyruvate kinase M$_1$, rabbit aldolase A, and papain (38), and also for the sensitive detection of amino acids in human serum and dried blood (39). Amino and imino acids in the biological specimens are extracted with ethanol and derivatized with NBD-fluoride as described above for 1 min. The method is about one order of magnitude more sensitive than the corresponding method using derivatization with o-phthalaldehyde. The amino acid contents obtained by the proposed method were comparable to those obtained by the amino acid analyzer using o-phthalaldehyde.

Postcolumn derivatization with NBD-fluoride and fluorometric detection were used for an automatic amino analyzer, by means of which the amino acid profiles in blood samples were analyzed (38).

Derivatization with the analogous reagent 7-chloro-4-nitrobenzo-2-oxa-1,3-diazole (NBD-chloride) is applied for the detection of picomole amounts of primary and secondary amino groups, especially of proline and hydroxyproline. Ahnoff et al. (40) describe the determination of 4-hydroxyproline in collagen derivatives, which is based on precolumn derivatization with NBD-chloride and separation on a LiChrosorb RP-18 column. The derivatives are eluted with phosphate buffer (pH 1.9), containing 20% acetonitrile and 5 mol m$^{-3}$ sodium heptanesulfonate. Recovery relative to that for pure standard solution is better than 95%.

Postcolumn derivatization with NBD-chloride is described by Yoshida et al. (41). The column eluate is mixed with borate buffer (pH 10.5) and reacted with an ethanolic solution of NBD-chloride. After the reaction, the resulting solution is mixed with equal volumes of an organic solvent, for example, alcohol, which contains 1 mol dm$^{-3}$ HCl. The optimum pH is 8.0–8.5; the optimum temperature range is 60–70°C.

A new method for the determination of primary and secondary amino acids is presented by Einarsson et al. (42). The procedure includes derivatization of the amino acids with 9-fluorenylmethyl chloroformate, followed by extraction with pentane. The derivatives are highly fluorescent and stable, with the exception of the histidine derivative, which shows any breakdown. The derivatives of 20 amino acids were separated by HPLC on a 3-$\mu$m Shandon ODS Hypersil column with a linear gradient from acetonitrile-methanol-acetate buffer (10 : 40 : 50) to acetonitrile-acetate buffer (50 : 50). The detection limits are in the low femtomole range. The method is applicable to protein hydrolysates, cerebrospinal fluids, blood serum, and urine with satisfactory results.

Protein microsequencing can be performed with a novel class of phenylisocyanates (43). These compounds are substituted in the 4-position of the phenyl ring with a protected amine and have the general structure *tert*-Boc-NH-(CH$_2$)$_n$-PITC [Boc = butoxycarbonyl, PITC = phenyl(isothiocyanate)]. Their most important feature is the generation of a primary amine during the acid cleavage step of the sequencing process (Fig. 5.2). This primary amine is then available

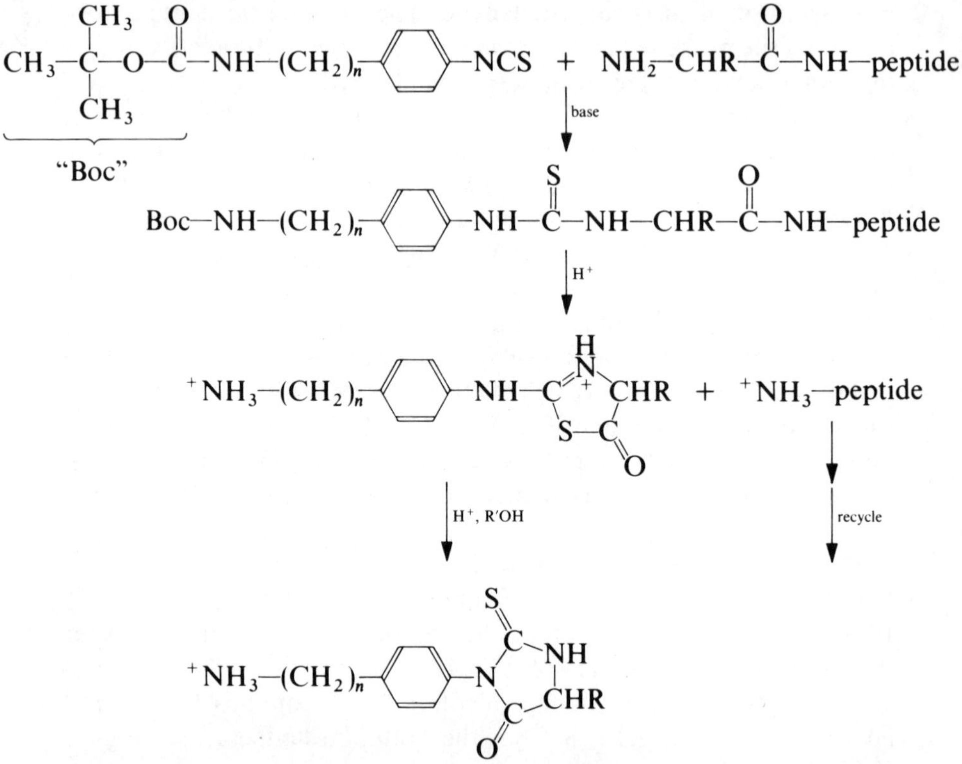

**Figure 5.2.** General scheme of the Edman degradation using the modified PITC reagent.

for fluorescent labeling at the time of identification to increase the detectability of the modified phenylthiohydantoin (PTH)-amino acid. Preliminary studies on the fluorescence detection of the aminomethyl-PTH-amino acids showed at least a 2.5-fold increase in sensitivity over the most commonly used detection method, UV detection at 254 nm. This approach may provide a general method for picomole and subpicomole protein-sequence determinations with currently available instrumentation.

## 1.2. Vitamins

During recent years, numerous HPLC/fluorometry methods for the determination of vitamins in biological, pharmaceutical, and food samples have been published. Table 5.2 shows some selected applications that are discussed here and in Section 3.

Table 5.2. HPLC Methods for the Determination of Vitamins

| Vitamin | Sample | Column Material | Mobile Phase | Section in Chapter 5 | References |
|---|---|---|---|---|---|
| A (retinol) | Serum | CN or silica gel | $n$-Hexane/iso-PrOH | 1.2 | 44, 45 |
| | Plasma | $C_{18}$ | Methanol | 1.2 | 44, 45 |
| $B_1$ (thiamin) | Rice flour | $C_{18}$ | $NaH_2PO_4/NaClO_4$ | 3 | 131 |
| $B_6$ compounds | Human milk | $C_{18}$ | Gradient | 1.2 | 46 |
| C (ascorbic acid, dehydroascorbic acid) | Whole blood | $C_{18}$ | $KH_2PO_4$/methanol | 1.2 | 47 |
| | Foodstuffs, beverages | $C_{18}$ | $KH_2PO_4$/methanol | 3 | 136 |
| | Foodstuffs | $C_{18}$ | Methanol/water and ion-pairing reagent | 3 | 132 |
| | Food | Anion exchange | — | 3 | 133 |
| $K_1, K_2$ | Human plasma | $C_8$ | Methanol (92.5%) | 1.2 | 48 |
| $K_3$ | Animal feed, premixes | $C_{18}$ | $H_2O$/methanol | 3 | 136 |

The determination of vitamin A (retinol) in blood serum was described by Biesalski et al. (44). After addition of 0.2 cm$^3$ acetonitrile to the serum and extraction with $n$-hexane, HPLC was performed on CN or silica gel columns with $n$-hexane-isopropanol (95 : 5) as the mobile phase. The recovery rate of retinol with this method is 97.3–101%.

Another HPLC method for the determination of retinol and its acetate in plasma samples was reported by Collins and Chow (45). The method employs a $C_{18}$ reversed-phase column and methanol as an eluant, and the detection is monitored with fluorescence excitation at 348 nm and emission at 470 nm. Detector noise established the lower limit of quantitation at about 0.5 ng. Vitamin A in plasma (as little as 1 mm$^3$) was quantitated at < 1 ng by this procedure.

A semiautomated and sensitive method for the fluorometric determination of vitamin $B_6$ (pyridoxal-5'-phosphate) in whole blood was described by Schrijver et al. (49). After the acid extraction of vitamin $B_6$, a fully automated HPLC system was used to separate it from interfering compounds. Vitamin $B_6$ was derivatized to its semicarbazone and detected fluorometrically. In routine analyses, 125 samples were run within 48 h. The within-assay and between-assay relative standard deviations are 3.3% and 4.7%, respectively.

Total vitamin C in whole blood can be determined with a semiautomated HPLC method described by Speek et al. (47). After deproteinization of whole blood, L-ascorbic acid is oxidized enzymatically to dehydro-L-ascorbic acid, and the latter is condensed with 1,2-phenylenediamine to its quinoxaline derivative (see Fig. 5.3). This derivative is separated on a reversed-phase ODS-Hypersil HPLC

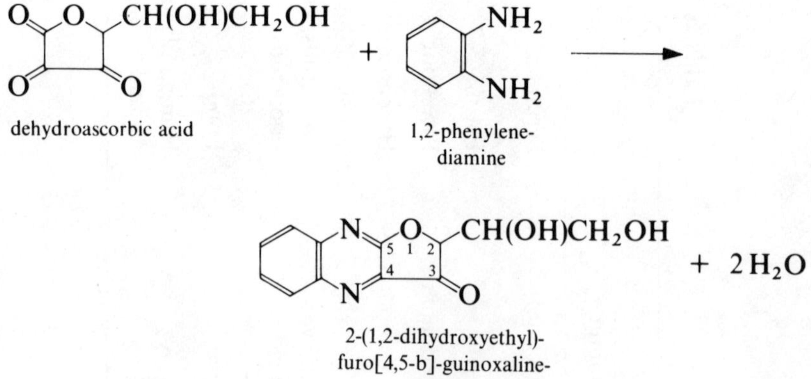

**Figure 5.3.** Derivatization of dehydroascorbic acid.

column, eluted isocratically with $KH_2PO_4$/methanol solution (pH 7.8) that was flushed with helium gas before use, and after the separation detected fluorometrically ($\lambda_{ex} = 365$ nm; $\lambda_{em} = 418$ nm). The quinoxaline derivative of vitamin C in the blood extract is stable for at least 24 h if stored in the dark at 4°C. Routine vitamin C determinations can be carried out in a series of 100 samples within 48 h. The within-assay and between-assay relative standard deviations were 3.7% and 4.6%, respectively. Concentrations as low as 0.2 mmol m$^{-3}$ of total vitamin C in whole blood could be detected by this method.

The simultaneous determination of ascorbic acid and dehydroascorbic acid after their HPLC separation can be performed by omitting the oxidation step (see Section 3).

A highly sensitive method for the fluorometric detection of $K_1$ and $K_2$ vitamins by using postcolumn electrochemical reduction is described by Langenberg and Tjaden (48). The K vitamins are isolated from plasma samples by extraction with $n$-hexane and separated by HPLC on a reversed-phase column. As K vitamins do not exhibit native fluorescence, they are reduced in a postcolumn reaction to highly fluorescent hydroquinone derivatives. Dual electrochemical detection in the reduction/reoxidation mode for coulometric/coulometric, as well as coulometric/amperometric, detection appears to be more sensitive and selective toward the plasma background than simple reductive electrochemical detection, but fluorometric detection after coulometric reduction offers the best results. Combination of normal-phase chromatography and the described method is possible only if supporting electrolyte is added to postcolumn, but leads to higher detection limits. The method was applied to the determination of vitamin $K_1$ in human plasma samples.

## 1.3. Steroids

### 1.3.1. Bile Acids

A bile acid consists of a steroid skeleton with a side chain at $C_{17}$, terminating in a carboxyl group (see Fig. 5.4). The bile acids differ in the number and position of the OH groups, which are all of the α-configuration (this means *trans* configuration to the $CH_3$ group at $C_{10}$). The structures of two important bile acids are shown in Fig. 5.4.

During recent years, numerous methods for the determination of bile acids in biological materials such as plasma, urine, and bile have been published. A procedure for the simultaneous determination of

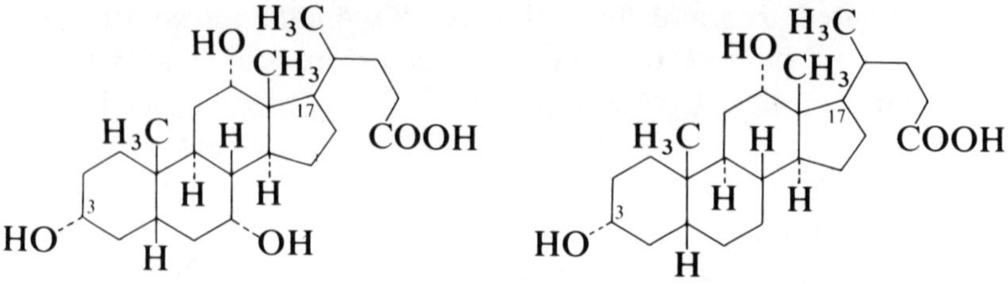

**Figure 5.4.** Structures of the bile acids, (a) cholic acid, and (b) deoxycholic acid.

bile acids in blood serum, involving precolumn derivatization with 1-anthroylnitrile, is described in detail in Section 2.11; other applications, which are all based on an enzymatic reaction, are also outlined in this chapter.

Bile acids, which are 3-hydroxy steroids, react with 3-hydroxy steroid dehydrogenase in the presence of NAD (nicotinamide-adenine dinucleotide). The hydroxy group at $C_3$ is oxidized to a keto group, while NAD is reduced to NADH, which is measured fluorometrically (50). The following applications are modifications of this reaction.

1. Bile acids were extracted from freeze-dried rat feces by refluxing in methanol/chloroform at 70°C for 10 h and were then separated by HPLC. Elution was performed with acetonitrile (10 mol m$^{-3}$)-KH$_2$PO$_4$ (2 : 3) and acetonitrile (30 mol m$^{-3}$)-KH$_2$PO$_4$ (1 : 4) gradients. The detection limit was about 10 ng (51).

2. The determination of bile acids in rat bile was presented by Imai et al. (52). A mixture of 10 mm$^3$ bile was deproteinized with ethanol and hydrolyzed with NaOH. Reversed-phase HPLC on a $C_8$ column was performed at 40°C by gradient elution in three steps. The 3-hydroxy steroid dehydrogenase was fixed on aminopropyl controlled pore glass in a stainless steel column at 20°C. Elution was completed within 40 min.

3. Free and conjugated bile acids in human sera were determined by micro HPLC (53). The bile acids were collected on a precolumn, packed with (octadecylsilane)-silica, by passing a diluted phosphate solution of a serum through it, and the separation was performed on a silica ODS SC-01 column. The detection limit was 0.13–0.28 pmol for a signal-to-noise ratio of 2. An amount of 0.1 cm$^3$ of serum was enough for determining each bile acid in it.

4. Takeda et al. (54) report a HPLC method for the analysis of bile acids in serum, which involves an automated pretreatment system with a reversed-phase column, line filter, values for injector, degasing apparatus, sampler, and pump. The interfering substances for the determination of bile acids by HPLC, including water-soluble proteins, peptides, and ions, could be eliminated within 2–5 min. The recoveries of cholic acid, glycocholic acid, and taurocholic acid added to serum were 100%, 101%, and 102%, respectively.

5. Bile acids in 25-mm$^3$ serum samples were determined by an automatic method, described by Hasegawa et al. (55). The procedure involves removal of most proteins and other hydrophilic compounds from serum by pretreatment, using a column with tetraethyleneglycol diacrylate-tetramethylolmethane triacrylate copolymer, then separation of the bile acids by reversed-phase HPLC on a larger column of the same polymer, with gradient elution by acetonitrile-$(NH_4)_3PO_4$ solution containing NAD, and the reaction of the separated bile acids on a column containing cellulose-immobilized 3-hydroxy steroid dehydrogenase. The determination takes about 1 h, and the standard curves for 15 bile acids and conjugates are linear up to 10 mmol m$^{-3}$, with sensitivities of 0.042–0.116 mmol m$^{-3}$.

6. Thirteen bile acids were determined by Karatani and Oka (56). An immobilized column was packed with partially fluorinated gel-like glass particles fixed with 3α-hydroxy steroid dehydrogenase as immobilized enzyme. Calibration curves were linear in the range of 2–500 ng. The detection limit was 1 ng, and the relative standard deviation was < 5%. In spite of repeated measurement of over 300 times of 500 ng bile acids, no change in the detection sensitivity or the degradation of activity of immobilized enzyme was observed.

### 1.3.2. Other Steroids

Different steroid derivatization methods with automatic, air-segmented reaction systems after HPLC separation are discussed by Reh and Schwedt (57). Derivatization with isonicotinoyl hydrazide (with chromatography on a Nucleosil-$NO_2$ column and elution with water-acetonitrile in carbon tetrachloride) or with periodate/acetylacetone (with chromatography on a LiChrosorb RP-8 column and elution with acetonitrile-water) is suitable. Detection limits for corticosterone were 100 and 7 ng with periodate/acetylacetone and isonicotinoyl hydrazide, respectively. The determination of urinary steroids after chromatography on Nucleosil-$NO_2$ with detection by UV spectrophotometry and fluorescence detection with isonicotinoyl hydrazide derivatization demonstrated improved sensitivity with the latter method.

Corticosterone, cortisol, and androstenedione were determined by Schwedt and Reh (58), coupling HPLC separation with postcolumn derivatization for subsequent fluorometric detection. Reversed-phase chromatography and adsorption chromatography, systems with chemical bonded phase, as well as a one-step reaction with isonicotinoyl hydrazide in organic solvents and a multistep reaction of oxosteroids to form dihydrolutidine derivatives via the Hantzsch reaction were combined. The reaction limits for corticosterone were 7–10 ng.

The determination of 17-oxosteroids in serum and urine is described by Kawasaki et al. (59). The steroids are extracted with dichloromethane after enzymic hydrolysis ($\beta$-glucuronidase-sulfatase), and dehydroepiandrosterone sulfate in serum samples is solvolyzed with sulfuric acid in ethyl acetate. The 17-oxosteroids are labeled with dansyl hydrazine in trichloroacetic acid-benzene solution and then chromatographed on a microparticulate silica gel column, using dichloromethane-ethanol-water (400 : 1 : 2) as the mobile phase. The eluate is monitored by a fluorophotometer at 365 nm (excitation) and 505 mm (emission). Linearities of the fluorescence intensities (peak heights) with the amounts of various 17-oxosteroids were obtained between 60 and 1000 pg.

Kawasaki et al. (60) also report a procedure for the direct determination of 17-oxosteroids in biological fluids without hydrolysis. The conjugated steroids are extracted with a $C_{18}$ cartridge, labeled again with dansyl hydrazine in trichloroacetic acid-benzene solution, and then separated by HPLC on a reversed-phase $C_{18}$ column, using sodium acetate in a methanol-water-acetic acid mixture as the mobile phase. The eluate is monitored as described above. Linearities were obtained between 10 and 100 pmol. The method is useful for the simultaneous determination of conjugated 17-oxosteroids in urine and serum.

17-Hydroxy corticosteroids in human biological fluids were separated by HPLC and derivatized in a postcolumn reaction with benzamidine (61). Fluorescence was measured at 480 nm, with excitation at 370 nm; the highest fluorescence intensity was achieved at pH 13–14. Limits of detection for cortisol were 5–50 ng added to the chromatographic column, and interassay relative standard deviations for tetrahydrocortisol and tetrahydrocortisone were 5.2% and 7.8%, respectively.

## 1.4. Organic Acids

Several determination methods consisting of precolumn derivatization, HPLC separation, and fluorometric detection have been reported. Three derivatization procedures for the precolumn fluorescence labeling of carboxylic acids were compared by Lingeman et al. (62). The methods are based on the formation of esters by coupling the acid with 9-(hydroxymethyl)anthracene (Fig. 5.5a). The carboxylic acid function is activated with 2-bromo-1-methylpyridinium iodide, $N,N'$-carbonyldiimidazole, or $N$-ethyl-$N'$-(3-dimethylaminopropyl)-carbodiimide hydrochloride. All three methods convert benzoic acid completely into the corresponding ester. After HPLC separation (column: LiChrosorb RP-18, mobile phase: methanol/water), about 100 fmol of the acid could be detected. The methods are well suited for the analysis of carboxylic acids in plasma.

For the precolumn derivatization of fatty acids, some fluorogenic reagents have been described that consist of a polyaromatic ring system.

1. The 12 biological most important fatty acids ($C_{12}$–$C_{22}$) were determined in samples (e.g., bile, blood serum, adipose tissue, food) by extraction with chloroform-methanol, total lipids fractionation by

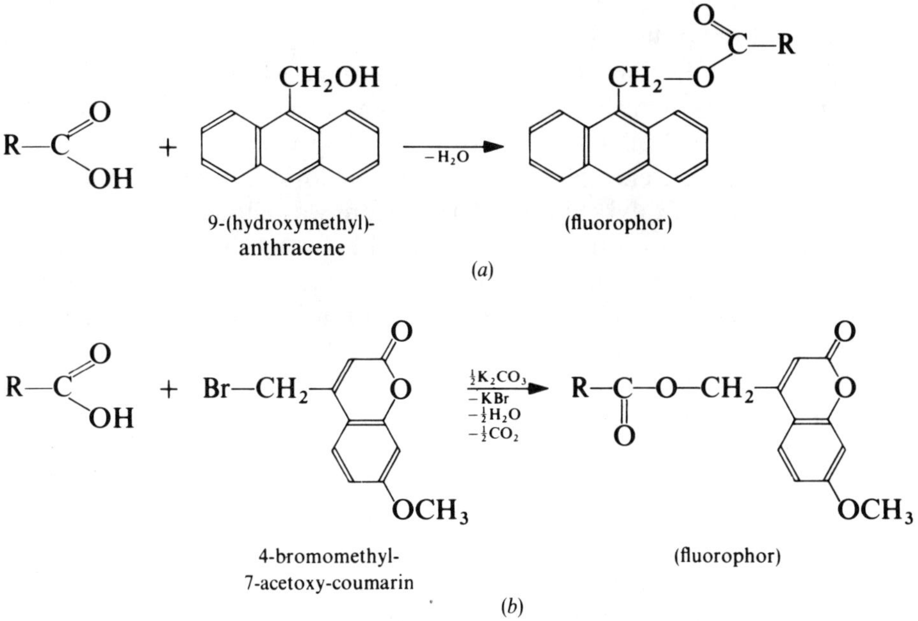

**Figure 5.5.** Derivatizations of organic acids.

thin-layer chromatography, hydrolysis, and treatment of the free fatty acids with 9-anthryldiazomethane to form fluorescent derivatives (63). HPLC separation was performed on two reversed-phase columns (Zorbax $C_8$ and LiChrosorb RP-8) at 35°C with elution by acetonitrile-water in the gradient mode. The derivatives were determined fluorometrically with excitation and emission wavelengths of 365 and 412 nm, respectively. The chromatographic separation required only 62 min, and complete separations between $C_{18}^{3=}$ and $C_{22}^{6=}$ and among $C_{14}^{0=}$, $C_{20}^{4=}$, and $C_{16}^{1=}$ were obtained. There was good linearity for each fatty acid in the range 10–1000 ng.

2. The simultaneous separation and sensitive determination of free fatty acids in blood plasma is described by Tsuchiya et al. (64). The fatty acids are derivatized with the fluorescent reagent 4-bromomethyl-7-acetoxycoumarin (Fig. 5.5b) and separated on a LiChrosorb RP-18 column. The eluted derivatives are hydrolyzed by mixing them with an alkaline solution, and the fluorescence produced is detected. The derivatives of series of both saturated and unsaturated fatty acids ($C_6^{0=}$–$C_{20}^{4=}$) are simultaneously separated by a continuous gradient elution method with a methanol-based solvent containing acetonitrile. This method is applicable to the quantitative determination of free fatty acid in human blood samples and in very low concentrations. Ten cubic millimeters of blood plasma is sufficient, and the results show good recovery and good reproducibility.

3. The application of 9-aminophenanthrene as a fluorescence reagent for the labeling of free fatty acids in human serum was examined by Ikeda et al. (65). The method involves dissolving the reagent in benzene to a benzene solution of the acid chlorides derived from the fatty acids and oxalyl chloride. The mixture was allowed to react for 45 min at 70°C to form derivatives with a strong fluorescence. The compounds thus obtained have wavelength maxima at around 303 nm for excitation and 376 nm for emission. By using this derivatization method, recoveries were measured for seven kinds of free fatty acids added to 0.5 cm³ of healthy human serum. Significant recoveries ranging from 96% to 107% (relative S.D. 1.4–5.0%) were obtained. Detection limits of free fatty acids by this derivatization method were 10 pmol for $C_{14}^{0=}$, $C_{16}^{0=}$, $C_{16}^{1=}$, $C_{18}^{1=}$, and $C_{18}^{2=}$, and 15 pmol for $C_{18}^{0=}$ and $C_{20}^{4=}$.

A HPLC method for the determination of uronic acids based on precolumn dansylation was developed by Takeda (66). Various uronic acids, such as glucuronic acid, galacturonic acid, and mannuronic acid, were mixed with a trichloroacetic acid-ethanol solution and

1.0% dansylhydrazine-ethanol solution. The mixture was incubated for 45 min at 40°C and then cooled to room temperature. An aliquot of the resulting solution was injected into a SIL-NH$_2$ column. Good separation of the uronic acids was obtained within 15 min by an elution system using acetonitrile/acetic acid buffer (pH 5.6). The fluorescence of the eluate was monitored at 530 nm with an excitation wavelength of 350 nm. Linearity of the fluorescence intensity with the amounts of uronic acids was obtained between 0.1 and 20 nmol. The detection limit was about 50 pmol of uronic acids.

The proposed method is applicable to the detection of various acids in glycosaminoglycans and other biological substances.

### 1.5. Other Compounds

In this section, some selected HPLC methods for the determination of compounds, which have not been covered in the previous sections, will be described.

#### 1.5.1. Alkaloids

Sasse et al. (67) describe the quantitation of the harmone alkaloids (Fig. 5.6) of *Peganum harmala* in cell culture extracts and also report a rapid fluorometric method for distinguishing between two groups of alkaloids in unpurified cell extracts. The different fluorescence spectra in methanol of harmine(I)/harmol(II) and harmaline(III)/harmalol(IV) allowed determination of the two groups of alkaloids in the part-per-billion range without separation. The fluorescence spectra of (III) and (IV) were not altered by dilution with water, whereas those of (I) and (II) were. The cell extracts were diluted with phosphate buffer (0.2 mol dm$^{-3}$, pH 5.0) to exclude errors caused by pH variations. The detection limit was $< 0.5$ $\mu$g cm$^{-3}$, and calibration graphs for (I) and (III) were linear to 1 and 3 $\mu$g cm$^{-3}$, respectively. In *Perganum* cultures

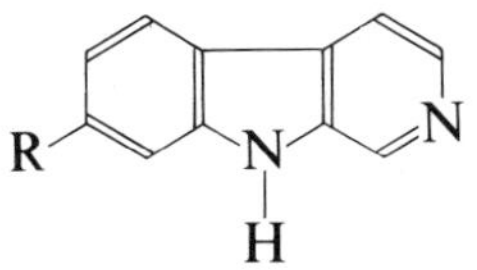

(fluorophor)   (fluorophor)

I; R = CH$_3$O (harmine)   III; R = CH$_3$O (harmaline)
II; R = OH (harmol)   IV; R = OH (harmalol)

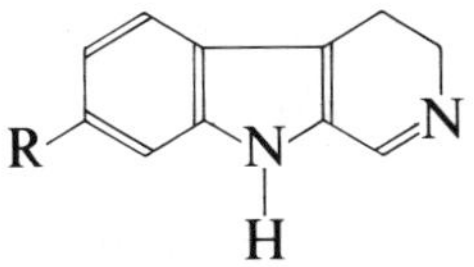

**Figure 5.6.** Structure of harmane alkaloids.

the concentration of (II) was only 5% that of (I), whereas (III) and (IV) were present in equal amounts. HPLC was performed on an RP-8 column combined with a RP-2 precolumn and isocratic elution with methanol-water-formic acid buffered with triethylamine at pH 8.5. All alkaloids could be detected by measuring the absorbance at 330 nm, but fluorometric detection increased the sensitivity 100-fold, and the detection limit was < 10 pg.

### 1.5.2. Thiols

The trace determination of biological thiols is described by Mopper and Delmas (68). In a reaction analogous to the derivatization of primary amines with *o*-phthalaldehyde and a thiol compound (see Fig. 4.2b), the thiols are converted to highly fluorescent isoindole derivatives by reaction with *o*-phthalaldehyde and an amino compound (2-aminoethanol). The derivatization is performed in aqueous solution at a mild pH and is completed within 1 min. The derivatives are separated on a $C_{18}$ reversed-phase column, using a gradient of two solutions (A: sodium acetate, pH 5.7; B: acetonitrile or methanol) as mobile phase. Unlike the situation with earlier precolumn derivatization techniques for thiols, there are no interfering reagent or reagent by-product peaks, a feature that is advantageous for ultratrace analysis of thiols in aqueous samples. The detection limit is about 25 fmol per injected thiol, and the precision is about $\pm$ 7% at the 2–3 pmol level. The linearity of response was examined over three orders of magnitude (nanomolar to micromolar) and was linear in that range. The method was used for the analysis of thiols in urine and the reducing marine sediment pore-waters.

Two other reagents for the precolumn derivatization of thiols were presented by Toyooka and Imai (69, 70).

1. The thiol compounds are derivatized with the fluorogenic reagent ammonium 7-fluorobenzo-2-oxa-1,3-diazole-4-sulfonate (SBD-F; Fig. 5.7a), separated on a HPLC column ($\mu$ Bondapak $C_{18}$) and detected fluorometrically at 515 nm with excitation at 385 nm. Two kinds of gradient system were adopted for the separation on SBD-cysteine, -homocysteine, -cysteamine, -glutathione, and -*N*-acetylcysteine. The detection limits were in the range 0.07–1.4 pmol. Only a reduced form of glutathione was found in human whole blood, at a level of 1.6 mol m$^{-3}$. However, in the plasma, both the reduced and oxidized (tri-*n*-butylphospine-treated) L-cysteine, glutathione, and an unknown substance were detected.

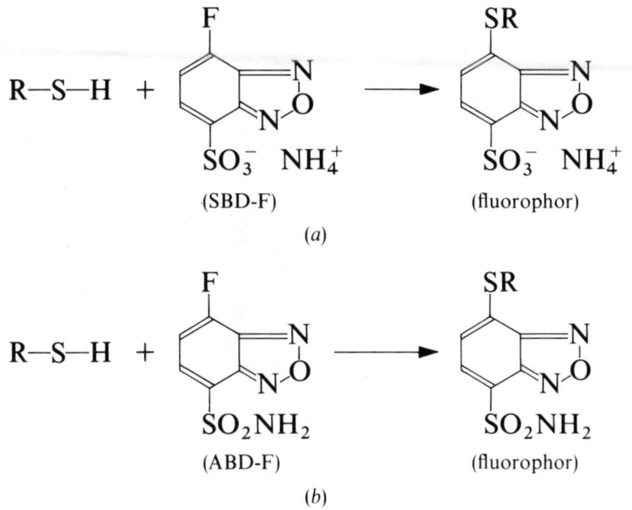

**Figure 5.7.** Derivatization of thiols.

2. The analogous compound 4-(aminosulfonyl)-7-fluorobenzo-2-oxa-1,3-diazole (ABD-F; Fig. 5.7b) can also be used as a reagent for thiols. The reaction rate of ABD-F with homocysteine is $> 30$ times faster than that of ammonium 7-fluorobenzo-2-oxa-1,3-diazole-4-sulfonate. The fluorogenic reaction with thiol is completed in 5 min at 50°C and pH 8.0. Alanine, proline, and cystine do not react under the same conditions. The fluorescence intensity of the fluorophor is pH dependent with the highest value at pH 2. The ABD-thiols obtained by the prelabeling technique were separated and detected by reversed-phase HPLC. The detection limits for cysteine, glutathione, N-acetylcysteine, homocysteine, and cysteamine were 0.6, 0.4, 1.9, 0.5, and 0.5 pmol, respectively.

## 2. BIOMEDICAL AND CLINICAL CHEMISTRY AND FISHERIES BIOCHEMISTRY

Recently, modern fluorescence analysis, coupled with HPLC, has been used increasingly for the separation and determination of some organic components in biomedical and clinical samples. It has an advantage over other analytical methods in that its specific resolution characteristics serve to differentiate some closely related chemicals.

In this section, we describe some interesting recent developments in the fields of biomedical and clinical chemistry and fisheries bio-

Table 5.3. HPLC Conditions

| Section in Chapter 5 | Stationary Phase | Mobile Phase | Flow Rate (cm$^3$ min$^{-1}$) | Wavelengths (nm) $\lambda_{ex}$ | $\lambda_{em}$ |
|---|---|---|---|---|---|
| 2.1 | Varian CN-10 Micropak | Sodium citrate-sodium chloride buffer (3.5 mol dm$^{-3}$, pH 4.68) | 1.67 | 325, 360 | 380 |
| 2.2 | $\mu$ Bondapak C$_{18}$ | Chloroform-iso-octane-methanol, gradient | 1.5 | Filter type 285 | Filtertype 340 |
| 2.3 | Hitachi gel No. 3042 | 0.02 mol dm$^{-3}$ KH$_2$PO$_4$ buffer (pH 3.7) or anhydrousmethanol-water (6 : 4) | 1.0 | 350 | Cutoff filter 505 |
| 2.4 | PXS 5/25 ODS | Dichloroethane-ethanol-water (948 : 35 : 17) | 1.5 | 220 | 608 |
| 2.5 | LiChrosorb RP-18 | Acetonitrile-water-1 mol dm$^{-3}$ KH$_2$PO$_4$ (300 : 700 : 0.05) | 0.6 or 1.1 | 280 | 370 |
| 2.6 | Spherisorb ODS | 0.1 mol dm$^{-3}$ Na$_2$HPO$_4$ buffer (pH 6)-methanol (9 : 1) | 2.0 | 200 | Cutoff filter 320–400 |
| 2.7 | | 2 mol m$^{-3}$ NaH$_2$PO$_4$ buffer containing 10 mol m$^{-3}$ sodium heptanesulfonate and 150 cm$^3$ dm$^{-3}$ acetonitrile | | | Filter type |

| | | | | |
|---|---|---|---|---|
| 2.8 | Hypersil ODS | Aqueous solution containing 0.3 mol dm$^{-3}$ KH$_2$PO$_4$ and 16.7% (vol/vol) methanol | 2.0 | 470 | 525 |
| 2.9 | LiChrosorb RP-18 | Acetonitrile-heptofluorobutyric acid-water (A) 10 : 0.1 : 89.9, (B) 50 : 0.1 : 49.9 and gradient of (A) and (B) | 1.5 | 322 | 415 |
| 2.10 | Zorbax ODS | Ammonium acetate buffer (12.5 mol m$^{-3}$, pH 4.0) | 1.5 | 294.4 Filter type | 405 Filter type |
| 2.11 | Cosmosil 5 C$_{18}$ | 0.3% K$_3$PO$_4$ buffer (pH 6) methanol (1 : 5) | 1.8 | 370 | 470 |
| 2.12 | Zorbax-C$_8$ | Water-methanol (1 : 9) | 1.3 | 345 | 416 |
| 2.13 | C$_{18}$ | Acetonitrile-phosphate buffer (pH 5.2) (34 : 66) | 1.5 | 280 | 308 |
| 2.14 | C$_{18}$ | Acetonitrile-phosphate buffer (20 mol m$^{-3}$, pH 5.2) (36 : 64) | 1.5 | 280 | 308 |
| 2.15 | Supelcosil LC-18 | Citric acid buffer (50 mol m$^{-3}$, pH 3.4)-methanol, gradient | 1.5 | Variable | Variable |
| 2.16 | Zorbax-NH$_2$ | Methanol-NaH$_2$PO$_4$ buffer (0.2 mol dm$^{-3}$, pH 3) (1 : 9) | 1.3 | 328 | 526 |

chemistry, using the above-described analytical technique. Table 5.3 summarizes the HPLC conditions in the cited papers.

### 2.1. Cerebrospinal Fluid Polyamine Monitoring in Central Nervous System Leukemia (71)

Rennert et al. (71) developed the HPLC method for the analysis of cerebrospinal fluid, utilizing the o-phthalaldehyde fluorescent-detection system, which allows the reproducible detection of acid-soluble polyamines in this extracellular fluid compartment (Fig. 5.8). Measurements made by this method in childhood acute lymphocytic leukemia appear to correlate with the remission-relapse status of the patient. The sensitivity of this analytic technique may allow much earlier detection of central nervous system disease. This technique is also useful to monitor the success of chemotherapy in treating leukemia patients.

### 2.2. Fluorescent Derivatives of Prostaglandins and Thromboxanes for Liquid Chromatography (72)

Fluorescent esters of the prostaglandins (PG) $D_2$, $E_2$, $F_{2\alpha}$, and 6-keto-$F_{1\alpha}$ and of thromboxane (TX) $B_2$ have been prepared by Turk et al. (72), using the reagent 4-bromomethyl-7-methoxycoumarin (Mmc-Br; Fig. 5.9) (73). All of these derivatives can be separated in a single run

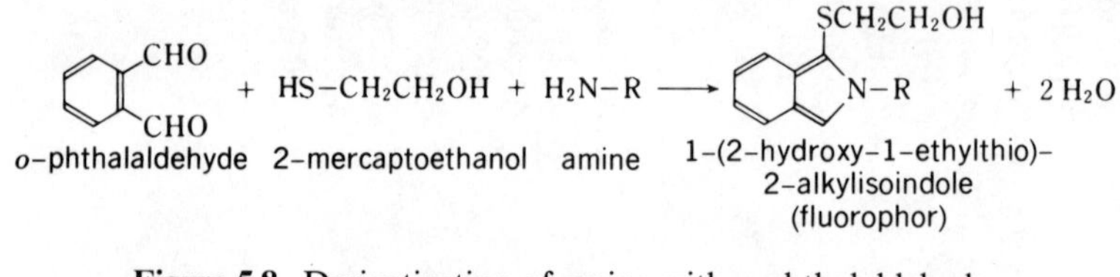

**Figure 5.8.** Derivatization of amine with o-phthalaldehyde.

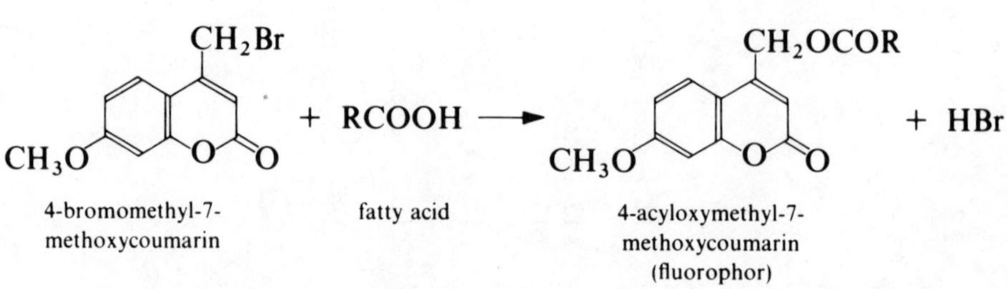

**Figure 5.9.** Derivatization of fatty acid with Mmc-Br as a fluorogenic reagent.

either by TLC or by HPLC. As little as 20 ng of $PGE_2$ can be detected after derivatization and HPLC analysis. Identification of $TXB_2$ produced by human platelets and of 6-keto-$PGF_{1\alpha}$ produced by bovine aortic microsome has been achieved with this method.

***Extraction Procedure:*** Biological preparations (74) containing prostaglandins were centrifuged for 10 min at 5,000 g to remove particulate matter and were then extracted with an equal volume of petroleum ether, which was discarded. From this point on, the handling of prostaglandin-containing aqueous solutions derived from biological and from chemical sources was identical. The solution was adjusted to pH 3.0 with 1 mol dm$^{-3}$ HCl and extracted three times with equal volumes of ethyl acetate. The pooled extracts were washed once with one-sixth volume of water to remove residual HCl, and the aqueous phase was discarded. The organic phase was dried over sodium sulfate, concentrated to dryness under nitrogen, and taken up in acetone.

***Derivatization Procedure***: Material to be derivatized was introduced into a 0.3-cm$^3$ (or 1.0-cm$^3$) reacti-vial as an acetone solution and concentrated to dryness under nitrogen. At least a threefold molar excess of Mmc-Br was then introduced as an acetone solution (2 mg cm$^{-3}$). A stirring bar and about 25 mg of potassium carbonate were then added, and the vials were capped and placed in a 67°C water bath for 10 min with magnetic stirring. The vials were

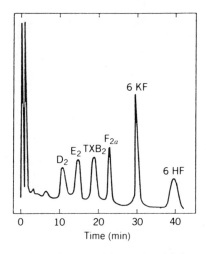

**Figure 5.10.** HPLC separation of Mmc adducts of multiple species. [Reprinted from J. Turk et al., *Prostaglandins*, **16**, 291 (1978).]

then placed on ice for about 5 min (to reduce solvent loss on opening), and an aliquot of the reaction solution was immediately subjected to chromatography.

An example of the result of HPLC is shown in Fig. 5.10.

### 2.3. Use of Native Fluorescence Measurements and the Stopped-Flow Scanning Technique in the HPLC Analysis of Catecholamines and Related Compounds (75)

Simultaneous separation of catecholamines and tryptophan (Fig. 5.11) metabolites has been carried out, using a reversed-phase partition mode of HPLC, by Krastulovic and Powell (75). Compounds are detected and measured via their native fluorescence emitted with an excitation wavelength of 285 nm and an emission cutoff filter of 340 nm. Sample preparation is minimized, and the assay is selective and well suited for routine analyses. The sensitivity of the method is in the nanogram range. The identity of chromatographic peaks is confirmed by their excitation spectra, obtained by the stopped-flow fluorescence scanning method.

This method is applied to the direct analysis of rat-brain and heart extracts as well as human serum samples.

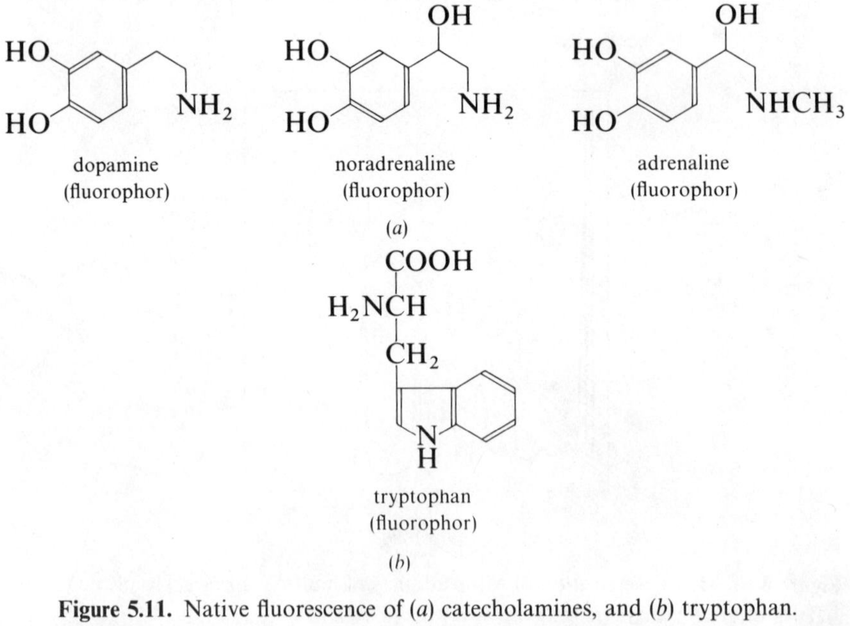

dopamine
(fluorophor)

noradrenaline
(fluorophor)

adrenaline
(fluorophor)

(a)

tryptophan
(fluorophor)

(b)

Figure 5.11. Native fluorescence of (a) catecholamines, and (b) tryptophan.

## 2.4. Determination of Plasma and Urinary Cortisol by HPLC Using Fluorescence Derivatization with Dansyl Hydrazine (76)

A method is described for the determination of cortisol in human plasma and urine by HPLC, using fluorophotometric detection, by Kawasaki et al. (76). After extraction with methylene chloride, cortisol is labeled with dansyl hydrazine (Fig. 5.12), and then separated by HPLC. The eluate is monitored by a fluorophotometer at 350 nm (excitation) and 505 nm (emission).

The optimum conditions for the determination, such as HCl and dansyl hydrazine concentrations, reaction time, and reaction temperature, and for the eluant of HPLC are discussed. Linearity of the fluorescence intensity (peak height) with the amount of cortisol was obtained between 0.5 and 60 ng. The recoveries for 50 and 100 ng of added cortisol were 98.7% and 95.4% for plasma, and 96.4% and 90% for urine, respectively. Comparision with a radioimmunoassay gave a correlation coefficient of .978. The proposed method is suitable for the routine analysis of cortisol in plasma and urine.

## 2.5. Determination of Urinary Placental Estriol by Reversed-Phase HPLC with Fluorescence Detection (77)

Taylor et al. (77) describe a HPLC procedure for determining urinary estriol (see Fig. 4.48) concentrations. The urine sample, after enzymatic hydrolysis to free the conjugated estrogen, is extracted with ether, and an aliquot of the resulting extraction residue is injected into the liquid chromatograph. Sample components are separated with a reversed-phase $C_{18}$ column and isocratic elution with an acetonitrile-water mobile phase. Using a far-UV excitation wavelength, the present authors measured the natural fluorescence of the eluted estrogen with a fluorescence detector. The procedure provides excellent sensitivity for determining near-term pregnancy concentrations of urinary estriol.

## 2.6. Simple and Rapid Determination of 5-Hydroxyindole-3-Acetic Acid in Urine by Direct Injection on a Liquid-Chromatographic Column (78)

5-Hydroxyindole-3-acetic acid (Fig. 5.13) has been determined by Wahlund and Kdlén (78) at normal and increased levels in urine. The urine samples were filtered or centrifuged and then injected into a liquid chromatograph containing a reversed-phase column with tri-

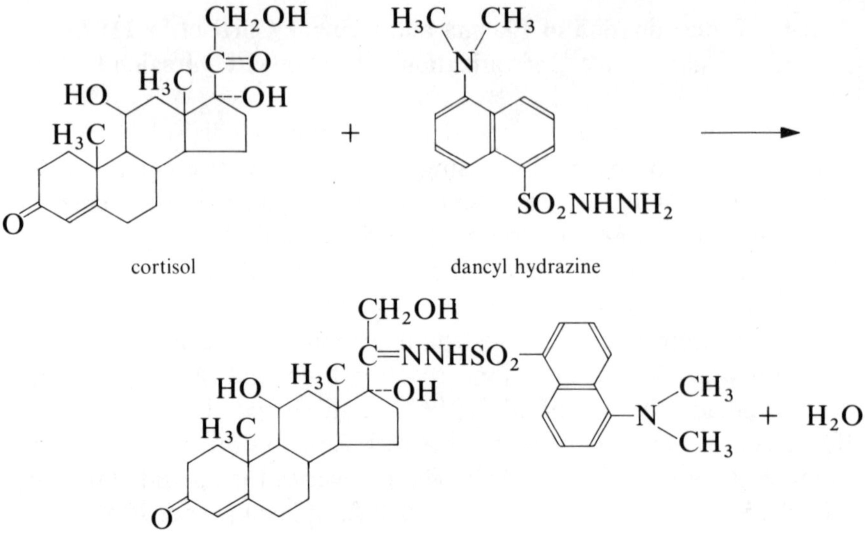

cortisol

dancyl hydrazine

dancyl hydrazonocortisol
(fluorophor)

**Figure 5.12.** Derivatization cortisol with hydrazine as an extrinsic fluorophor.

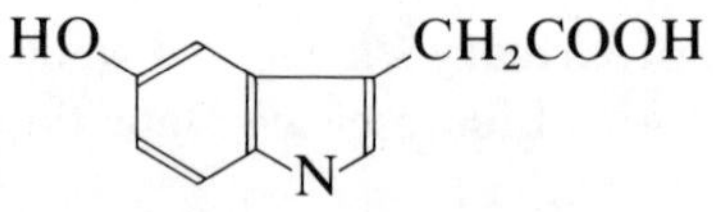

(fluorophor)

**Figure 5.13.** Structure of 5-hydroxyindole-3-acetic acid.

butyl phosphate as stationary liquid phase and an aqueous buffer and methanol as eluant.

5-Hydroxyindole-3-acetic acid is detected in the eluate by a fluorometric detector coupled to the outlet of the separation column, and the quantitation is performed by measurement of the peak heights of the sample and external standards. The percentage recovery was 97, and the precision 0.01. The method is suitable for the determination of increased levels of 5-hydroxyindole-3-acetic acid in urine from patients suffering from carcinoid tumours. Indole-3-acetic acid can be determined simultaneously.

### 2.7. Simple Method for the Assay of Urinary Metanephrines Using HPLC with Fluorescence Detection (79)

A method described by Jackman (79) extends a HPLC assay (80) for urinary catecholamines to the assay of urinary metanephrines (Fig. 5.14). The amines are separated from urine after acid hydrolysis

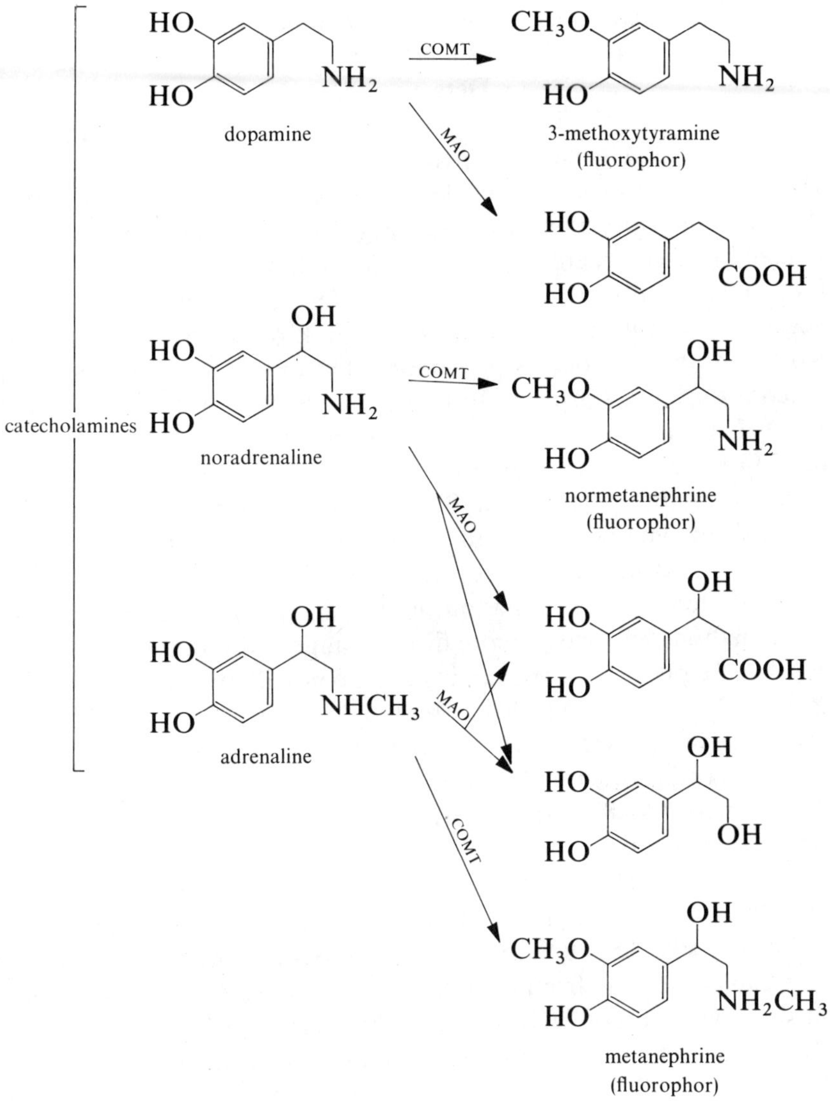

**Figure 5.14.** Metabolism of catecholamines.

COMT: catechol methyltransferase
MAO: amine oxidase (flavin-containing)

of conjugates by ion-exchange chromatography, and then further purified by solvent extraction. The final extracts are suitable for direct HPLC assay, using the endogenous fluorescence of the amines for detection. The 24-h excretion of the amines in 35 hospital in-patients was found to be (mean and range) as follows: 169 $\mu$g (24 h)$^{-1}$ normetanephrine (39–423), 102 $\mu$g (24 h)$^{-1}$ metanephrine (19–290), and 138 $\mu$g (24 h)$^{-1}$ 3-methoxytyramine (32–234).

## 2.8. Determination of the $B_2$ Vitamer Flavin-Adenine Dinucleotide in Whole Blood by HPLC with Fluorometric Detection (81)

Vitamin $B_2$ is a generic term for the three naturally occurring $B_2$ vitamers: flavin-adenine dinucleotide (FAD), flavin mononucleotide (FMN), and riboflavin (RB) (see Fig. 2.15) (82). The first two are phosphorylated, whereas the third is not. Vitamin $B_2$ occurs in various foodstuffs mainly as FAD and is widely distributed in all leafy vegetables and in meat, fish, eggs, and milk products (83). In the gut Rb taken up with food is phosphorylated to FMN by the intestinal mucosa during absorption. In tissue cells FMN can be converted to FAD. This vitamer serves as the coenzyme in most of the oxidation-reduction reactions catalyzed by the flavin enzymes.

Determination of vitamin $B_2$ in human blood is often used to establish the vitamin $B_2$ status of the individual (84). Various microbiological and manual fluorometric methods (82, 84–87) have been described for the determination of vitamin $B_2$ in blood, but each of these methods has one or more disadvantages. Microbiological methods (84) are rather time consuming and do not allow the analysis of the specific $B_2$ vitamers. Furthermore, they cannot be easily automated, and their results are poorly reproducible. Also, most of the fluorometric methods described do not allow separate analysis of the three $B_2$ vitamers (84–87), are laborious, and, like the microbiological methods, are not easy to automate.

Because of these drawbacks, Speek et al. (81) set out to develop a reliable HPLC method for the analysis of FAD in human whole blood. Their HPLC method was able to separate FAD, FMN, and Rb from each other and from interfering compounds. Also, a reliable and sensitive detection of FAD has been obtained by selecting the optimal pH of the mobile phase and optimal adjustment of the fluorescence detector. This PHLC method for the fluorometric analysis of FAD in whole blood, which is suitable for large-scale routine analysis, is now available.

## 2.9. Liquid-Chromatographic Profiling of Endogenous Fluorescent Substances in Sera and Urine of Uremic and Normal Subjects (88)

Several investigators have reported that an endogenous fluorescent substance increases in the blood of patients with chronic uremia

(89–92). This fluorescence is considered to be associated with albumin and is a source of error in the fluorometric assay of creatine kinase BB isoenzyme (89). Schwertner (92), using gel chromatography and TLC, isolated from uremic body fluids and normal urine a highly water-soluble, strongly fluorescent substance having an emission maximum of 415 ± 5 nm.

Mabuchi and Nakahashi (88) reported the use of HPLC to profile these fluorescent substances in both uremic and normal body fluids. The fluorescence excitation and emission maxima used by the present authors were 322 and 415 nm, respectively. Of the numerous fluorescent substances found in uremic body fluids and in normal urine (Fig. 5.15), some were also detectable in normal serum, but at relatively weak fluorescence intensities.

## 2.10. Liquid-Chromatographic Study of Fluorescent Materials in Uremic Fluids (93)

Using reversed-phase HPLC with fluorescence detection, Swan et al. (93) separated and identified some naturally fluorescent compounds in uremic serum and hemodialysate from patients with chronic renal disease. Several of the naturally fluorescent compounds were identified as indole derivatives by cochromatography with authentic standards. Compounds identified included indican, kynurenic acid, tryptophan, and 5-hydroxylindole-3-acetic acid (Fig. 5.16). Comparison of normal and uremic serum showed that the fluorescent materials are present in significantly greater concentrations in samples from uremic patients (Fig. 5.17).

## 2.11. Determination of Serum Bile Acids by HPLC with Fluorescence Labeling (94)

A method for the simultaneous determination of bile acids in serum by HPLC with fluorescence labeling is described by Gotô et al. (94). The bile acid fraction was obtained from a serum specimen by passing it through a Bond Elut cartridge. Bile acids were derivatized quantitatively into the fluorescent compounds through the hydroxyl group at C-3 by treatment with 1-anthroylnitrile in the presence of quinuclidine in acetonitrile (Fig. 5.18). These derivatives were separated into the free, glycine-conjugate, and taurine-conjugate fractions by ion-exchange chromatography on a lipophilic gel, piperidinohydroxypropyl Sephadex LH-20. Subsequent resolution of each fraction into cholate,

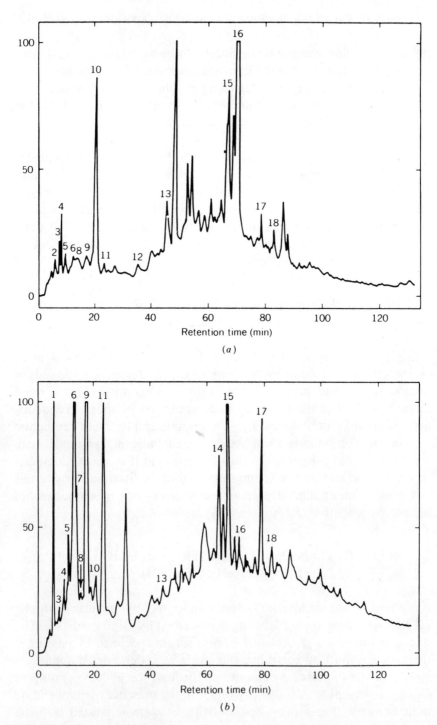

**Figure 5.15.** Typical chromatogram of normal urine (*a*) and uremic urine (*b*), representing an original urine volume of $2 \text{ cm}^3$. [Reprinted with permission from H. Mabuchi and H. Nakahashi, *Clin. Chem.*, **29**, 675 (1983).]

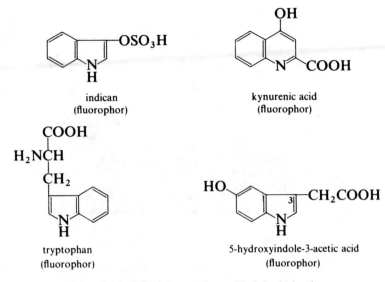

Figure 5.16. Native fluorophors of indole derivatives.

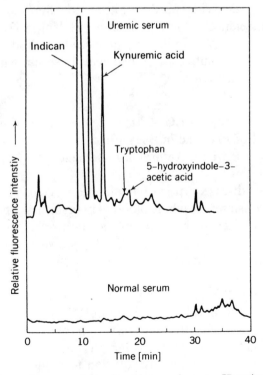

Figure 5.17. Chromatograms of uremic and normal serum. [Reprinted with permission from J. S. Swan, E. Y. Kragten, and H. Veening, *Clin. Chem.*, **29**, 1082 (1983).]

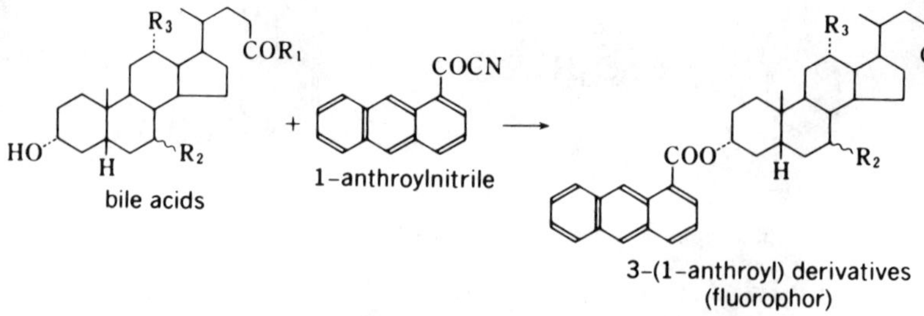


$R_1$ = OH, $NHCH_2COOH$, $NHCH_2CH_2SO_3H$, H
$R_2$ = H, $\alpha$-OH, $\beta$-OH
$R_3$ = H, $\alpha$-OH

**Figure 5.18.** Transformation of bile acids with 1-anthroylnitrile into the 3-(1-anthroyl) derivatives.

ursodeoxycholate, chenodeoxycholate, deoxycholate, and lithocholate was attained by HPLC on a Cosmosil $5C_{18}$ column, using 0.3% potassium phosphate buffer (pH 6.0)-methanol (1 : 5) and 0.1% potassium phosphate buffer (pH 6.0)-methanol (1 : 8) as mobile phases. The anthroyl bile acids were monitored by fluorescence detection (excitation wavelength 370 nm, emission wavelength 470 nm), the limit of detection being 20 fmol. The proposed method proved to be applicable to the quantitation of bile acids in serum with satisfactory reliability and sensitivity.

### 2.12. Fluorescence HPLC of Eicosapentaenoic Acid in Serum and Whole Blood of Fish and in Body Fluid of Plankton after Labeling with 9-Anthryldiazomethane (95)

A certain heart disease, myocardial infarction, occurs rarely in Eskimo people who consume the whole blood of many fishes. Recently, a main cause for this fact is believed medically to be the existence of 5,8,11,14,17-eicosapentaenoic acid (EPA) as a free unsaturated fatty acid in the vital blood of fish, which may prevent thrombosis in human blood vessels.

Ichinose et al. (95) have already reported on a fluorometric HPLC method using 9-anthryldiazomethane (ADAM) (96–100) as a fundamental experiment for the determination of EPA in some living bodies, such as those of human beings, fishes, and plankton (101). In the present work, a new microseparation technique for the purification of EPA from biological samples has been established. Part of the serum and vital blood of fish, diluted with water or the body fluid of plankton, was purified by passing through a micro-glass column packed with Extrelut using chloroform as the eluant (Fig. 5.19). After esterifying the desired constituent in the eluate with ADAM (Fig. 5.20), an aliquot of the ester was analyzed by the present reversed-phase HPLC method.

The proposed method for the analysis of EPA in the serum and vital blood of fish and in the body fluid of plankton is more sensitive and selective than the conventional gas-chromatographic methods (102–106) and can be applied also to the separation and determination of EPA in human serum and blood, as well as in fish and plankton.

An example of a chromatogram is shown in Fig. 5.21.

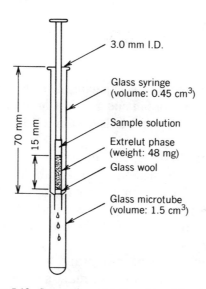

**Figure 5.19.** Separation of fatty acids with column.

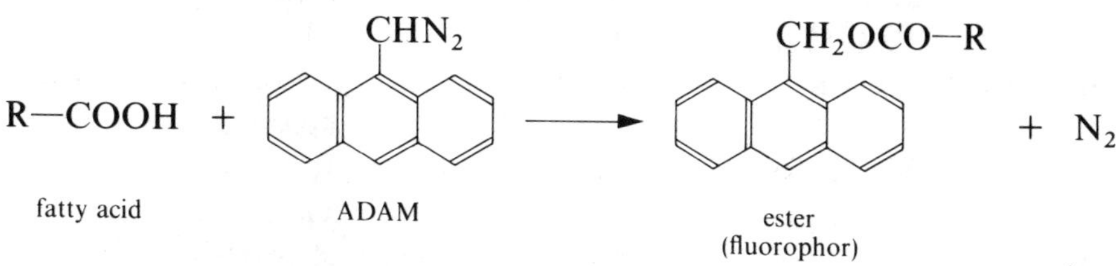

R—COOH + ADAM (CHN$_2$) ⟶ ester/fluorophor (CH$_2$OCO—R) + N$_2$

fatty acid         ADAM                    ester
                                       (fluorophor)

**Figure 5.20.** Esterification of fatty acids with ADAM.

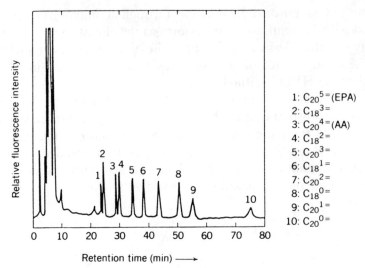

**Figure 5.21.** HPLC separation of the ADAM derivatives of $C_{18}$ and $C_{20}$ fatty acids.

### 2.13. Improved Assay of Unconjugated Estriol in Maternal Serum or Plasma by Adsorption and Liquid Chromatography with Fluorometric Detection (107)

Since Gurpide et al. (108) first elaborated an immunoassay for determining estrogens in biological liquids, many radioimmunoassay (RIA) methods (109, 110) have been proposed for monitoring estriol in maternal serum, and this technique is currently the most widely used. The ordinary fluorometric method for estrogens was mentioned in Chapter 4, Section 2.4.

Only a few years ago the only valid alternative technique for measuring estriol concentrations in maternal serum was gas chromatography with the use of a capillary column (111). However, this time-consuming technique involves many steps in sample preparation and requires highly specialized technicians.

During the course of the present experiments, two papers appeared reporting the determination of unconjugated estriol in pregnancy serum by HPLC (112, 113). Andreolini et al. (107) described a simple procedure involving HPLC with fluorometric detection, which has a sensitivity comparable to that of RIA methods. The sample of serum or plasma (500 mm³) is diluted 20-fold with water, and the estriol is absorbed from it onto graphitized carbon black (Carbopack B, Supelco) (114–116). After two washings the estriol is desorbed with chloroform/methanol (60/40 by volume), which then is evaporated.

The residue is redissolved in 50 mm$^3$ of water/acetonitrile, and 20 mm$^3$ is injected into the chromatograph.

The analytical recovery for estriol-supplemented serum or plasma averaged 98.6%. Day-to-day CVs ranged from 3.9% at 2 $\mu$g dm$^{-3}$ to 2.1% at 20 $\mu$g dm$^{-3}$. The limit of sensitivity is 0.3 $\mu$g dm$^{-3}$, which makes this procedure suitable for determination of estriol even in the first half of pregnancy. A single assay can be done within 30 min, 10 samples within 90 min. The present method is inexpensive, and shows that liquid chromatography can be used to determine estriol in pregnancy serum or plasma. It also is more sensitive and precise and requires less sample than other such methods.

### 2.14. Improved Determination of Estriol-16α-Glucuronide in Pregnancy Urine by Direct Liquid Chromatography with Fluorescence Detection (117)

Andreolini et al. (117) established a relatively simple, rapid assay of estriol-16α-glucuronide (Fig. 5.22) in pregnancy urine. The urine sample is diluted 20-fold with phosphate buffer (pH 5.2) containing 360 cm$^3$ dm$^{-3}$ of acetonitrile and 2 g dm$^{-3}$ of acetyltrimethylammonium bromide and is then directly injected into the chromatograph. A sample can be assayed within 14 min. Day-to-day CVs ranged from 2.3% at 45 mg dm$^{-3}$ to 2.9% at 4.8 mg dm$^{-3}$. The limit of sensitivity is 0.4 mg dm$^{-3}$. Results by the present method ($y$) correlated and compared very well with those by a method involving fractionation of estrogen conjugates and gas chromatography ($x$) for 24 samples of pregnancy urine ($y = 1.09x + 0.303$; $r = 0.947$). This assay is inexpensive and suitable for complete automation.

### 2.15. Liquid-Chromatographic Study of Fluorescent Compounds in Hemodialysate Solutions (118)

Barnett and Veening (118) have separated and identified three endogenous, naturally fluorescent substances in uremic hemodialysate by

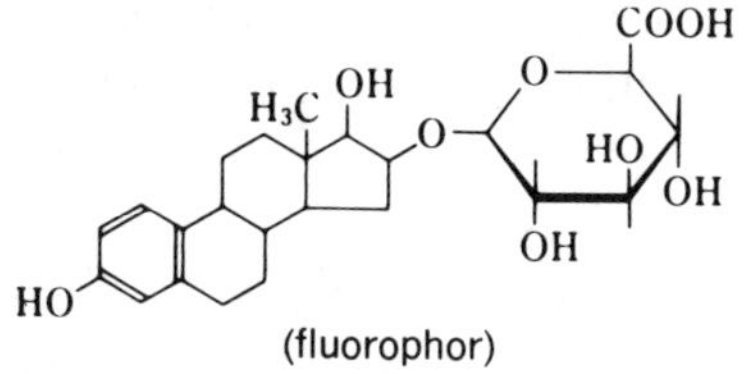

**Figure 5.22.** Structure of estriol-16α-glucuronide.

using reversed-phase liquid chromatography with fluorescence detection. Cochromatography with authentic standards, monitoring peak shifts after enzymic treatment, and spectrofluorescence measurements were used to confirm the identity of indican, tryptophan, and indole-3-acetic acid (see Fig. 5.16). Concentrations of indican were about 1.5 those of tryptophan and considerably greater than those of indole-3-acetic acid in hemodialysate samples from 12 renal patients. These three compounds, as well as eight unidentified components, were consistently present in dialysate samples from each of the 12 patients.

### 2.16. Determination of $B_2$ Vitamers in Serum of Fish Using HPLC with Fluorescence Detection (119)

$B_2$ vitamers consist of riboflavin (RF), flavin mononucleotide (FMN), and flavin adenine dinucleotide (FAD). FMN and FAD are protein bound, and the latter compound is usually the most abundant in natural products. It is well known that vitamin $B_2$ is a dietetically important compound, and many reports of its determination in various materials such as pharmaceutical products, food, beverages, and body tissues have been published.

In most determinations the sample is hydrolyzed by acid and enzyme in order to liberate FMN and FAD from the protein moiety in the sample with conversion into RF, which is subsequently determined spectrophotometrically or fluorometrically. Fluorometric methods are considered to be more sensitive, and measurements can be taken in either of two ways: by measuring the natural fluorescence of $B_2$ vitamers, or by measuring the fluorescence of lumiflavin (LF), which is derived from RF by irradiation with light, in order to enhance the fluorescence intensity.

HPLC has also been used for the separation and determination of RF and LF (120–124), especially for the analysis of pharmaceutical products containing multivitamins (125–127). However, HPLC has not been used for the separation and determination of RF, FMN, and FAD. The determination of FAD in human whole blood by HPLC in order to establish the normal concentration levels was carried out by Speek et al. (81) (see Section 2.8), but the development of this method for the determination of the other two $B_2$ vitamers has not been attempted.

Ichinose et al. (119) established the separation and determination of RF, FMN and FAD using HPLC with fluorescence detection and applied this method to the determination of low levels of $B_2$ vitamers in the serum of fish. Trichloroacetic acid was used to isolate $B_2$ vitamers from the serum. The detection limits of RF, FMN, and FAD in the serum were 4.89, 9.13, and 73.1 $ng\,cm^{-3}$, respectively.

## 3. FOOD CHEMISTRY

In Sections 1 and 2, several methods have been described that are based on HPLC procedures and fluorescence measurements and that allow the determination of relevant compounds in general biological and biomedical chemistry. In this section, applications in another field of interest, the analysis of food and feed samples, are presented. Some selected methods for the determination of amines, vitamins, and so on are summarized in Table 5.4 and will be explained in more detail in the following discussions.

The determination of volatile amines and ammonia in meat tissue is described by Parris (128). Amines and ammonia are determined by measuring the fluorescence of their dansyl derivatives (see Chapter 4, Section 1.1) after HPLC separation on a $C_{18}$ column with $KH_2PO_4$/acetonitrile as mobile phase. The procedure minimizes formation of both dansyl acid and dimethylamine from dansyl chloride and decomposition of dansylated α-amino acids to dansylamide. The determination of ammonia is more accurate and precise by this method than by the glutamate dehydrogenase assay.

The derivatization of α,ω-diamines (1,2-diaminoethane and its homologs up to 1,6-diaminohexane) to fluorescent bis(diphenylboron) chelates and their separation by HPLC are described by Claas et al. (129). In a precolumn derivatization step, the diamines are reacted with the reagent 2,2-diphenyl-1-oxa-3-oxania-2-borata-naphthalene (salicylaldehyde-diphenylboron chelate) to form the fluorescent products shown in Fig. 5.23, which are separated on a HPLC silica gel 60 column with a mixture of dichloromethane and n-hexane as mobile phase. Fluorescence is measured at 483–456 nm (emission) with an excitation wavelength of 365 nm. The smallest determinable concentration ranges from 0.3 ng for 1,2-diaminoethane to 17 ng for 1,6-diaminohexane dissolved in 50 $mm^3$ injected solution. 1,2-Diaminoethane, 1,4-diaminobutane, and 1,5-diaminopentane could be identified in samples of putrefied meat. The contents of 1,2-diaminoethane in 100 g of meat stored at room temperature increased the values from near zero to 200 μg within 8 days.

Predominant aliphatic amines in beer were determined by Murray and Sepaniak (130). The amines were derivatized with 7-chloro-4-

Table 5.4. HPLC Methods for the Determination of Biochemically Relevant Compounds in Food and Feed Samples

| Compound | Material | Method | Reference |
|---|---|---|---|
| Ammonia, volatile amines | Meet tissue | Derivatization by dansylation<br>HPLC: $C_{18}$ | 128 |
| $\alpha,\omega$-diamines | Putrefied meat | Precolumn derivatization to bis(diphenylboron)chelates<br>HPLC: silica gel 60 | 129 |
| Aliphatic amines | Beer | Precolumn derivatization with NBD-chloride<br>HPLC: $C_{18}$ | 130 |
| Vitamin $B_1$ (thiamin) | Rice flour | Postcolumn derivatization to thiochrome<br>HPLC: $C_{18}$ | 131 |
| Vitamin C (ascorbic acid, dehydroascorbic acid) | Foodstuffs | Precolumn derivatization with 1,2-phenylenediamine<br>HPLC: $C_{18}$ | 132 |
| | Foods, beverages | HPLC: $C_{18}$ | 134 |
| | Foods | HPLC: anion exchange | 133 |
| Vitamin E (tocopherols) | Seed oils | Native fluorescence | 135 |
| Vitamin $K_3$ (menadione sodium bisulfite) | Animal feed | Pre-column derivatization to menadione; post-column reduction<br>HPLC: $C_{18}$ | 136 |
| Indoleacetic acid | Radish plant | Native fluorescence<br>HPLC: $C_{18}$ | 137 |
| Mildiomycin | Cucumber, tobacco leaves | Precolumn derivatization with fluorescamine<br>HPLC: ion-pair/RP | 138 |

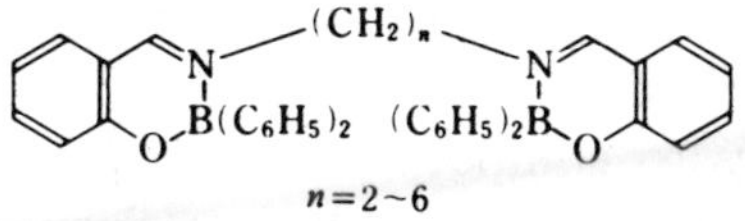

**Figure 5.23.** Fluorescent derivatization product of α,ω-diamines.

nitrobenzo-2-oxa-1,3-diazole (NBD-chloride; see Fig. 4.2e), and the derivatives were separated on a reversed-phase $C_{18}$ column using an acetonitrile-water solvent gradient. Linearity was achieved over four decades of concentrations with detection limits at about 5 pg.

Ohta et al. (131) report the determination of vitamin $B_1$ (thiamin) in brown and polished rice flour. HPLC separation was performed on a Nucleosil $C_{18}$ column with $NaH_2PO_4$ and $NaClO_4$ solutions as mobile phase. Thiamin was derivatized in a postcolumn reaction by addition of 0.1% $H_3Fe(CN)_6$ and 12% NaOH to produce thiochrome (see Fig. 4.11). Of 0.2 mg thiamin added to 100 g brown rice containing 0.46 mg/100 g, 95% (0.65 mg/100 g) was recovered.

In Section 1.2, the determination of total vitamin C (ascorbic acid and dehydro-L-ascorbic acid) in whole blood by oxidation and reaction with 1,2-phenylendiamine was described. By omission of the oxidation step, dehydroascorbic acid can be determined in the presence of ascorbic acid. This reaction principle was utilized by Keating and Haddad (132) for simultaneous determination of ascorbic acid and dehydroascorbic acid in foodstuffs. After precolumn derivatization of dehydroascorbic acid with 1,2-phenylendiamine, reversed-phase ion-pair HPLC is performed on a $C_{18}$ column with a mobile phase of methanol/water containing hexadecyltrimethyl ammonium bromide or tridecylammonium formate, respectively. The detector wavelength was initially set at 348 nm until the quinoxaline derivative of dehydroascorbic acid had eluted, and was then changed to 290 nm for the detection of ascorbic acid.

The same derivatization is also utilized in a postcolumn reaction for the determination of ascorbic acid and dehydroascorbic acid in complex matrices (133). The HPLC separation is accomplished with an anion-exchange resin, and fluorescence detection is achieved by oxidation of ascorbic acid to dehydroascorbic acid, followed by reaction with o-phenylenediamine. The lower limits of detection for both forms of vitamin C are well below the levels found in the usual food sources of this vitamin. A variety of foods, including fruit juices, vegetables, and fruits, were analyzed.

The simultaneous determination of total vitamin C and its C-5 epimer, total erythorbic acid (isovitamin C), in foods and beverages is described by Speek et al. (134). Vitamin C and isovitamin C are extracted and oxidized enzymically by ascorbate oxidase to the corresponding dehydro compounds. These products are condensed with $o$-phenylenediamine as described in Section 1.2 to the highly fluorescent quinoxaline derivatives. Then derivatives are separated on a $C_{18}$ column with a mobile phase containing $KH_2PO_4$ and methanol and detected fluorometrically at $\lambda_{ex} = 355$ and $\lambda_{em} = 425$ nm. Total vitamin C and isovitamin C can be detected at $> 0.2\ \mu g\,g^{-1}$. The amounts of dehydroascorbic acid and dehydroerythorbic acid present in foods and beverages can be determined by the same procedure by omission of the enzymic oxidation.

Speek et al. (135) report a HPLC method for the simultaneous analysis of E vitamins $\alpha$-,$\beta$-,$\gamma$-, and $\delta$-tocopherol and $\alpha$-tocotrienol in seed oils. After diluting the oils with $n$-hexane, the E vitamins are separated by HPLC and detected fluorometrically. Standardization was achieved, using electron-impact mass spectrometry and HPLC. The vitamin E compounds in several hot- and cold-pressed seed oils, originating from maize germs, olives, and soybeans and from sesame and safflower seeds, were investigated, and no differences were observed between E-vitamin concentrations of hot- and cold-pressed oils of the same origin. The vitamin E composition of oils of different origin varied widely. Of the oils examined, only maize germ oil contained $\alpha$-tocotrienol in detectable amounts (about 2%). Esterified E vitamin were not detected.

A HPLC method for the determination of vitamin $K_3$ (menadione sodium bisulfite) in animal feed and premixes is described by Speek et al. (136). After extraction the vitamin is converted into menadione, which is extracted with $n$-hexane and separated on a reversed-phase HPLC column (ODS-Hypersil) with a mobile phase of He-purged water-methanol. In a postcolumn reaction coil, menadione is reduced to 2-methyl-1,4-dihydroxynapthalene (Fig. 5.24), which is detected by fluorometric measurement. Several types of animal feed and premixes were analyzed according to this method, and concentrations of vitamin $K_3$ as low as 0.02 $\mu g\,g^{-1}$ could be detected. The within-assay

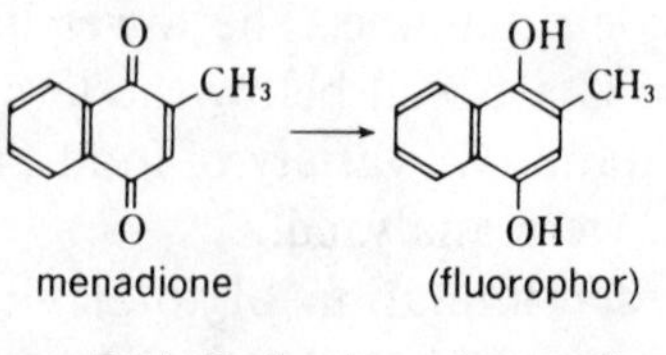

**Figure 5.24.** Reduction of menadione.

relative standard deviation of the method applied to feeds is 6.0%. The within-assay recovery of menadione Na bisulfite added to feeds is 94.4%.

A simplified method for the quantitative determination of indoleacetic acid in radish plants extracts is reported by Akiyama et al. (137). A C-18 SEP-PAK cartridge was used for the purification of the indoleacetic acid, and the separation was carried out by reversed-phase HPLC on a $C_{18}$ column. The endogenous indoleacetic acid content in green radish seedlings, corrected by internal standard methods with indolepropionic acid, was 20.2 ng g$^{-1}$ fresh weight.

A HPLC method for the determination of mildiomycin residues in plants and soils was established by Inoue and Hagimoto (138). Mildiomycin is extracted from cucumber and tobacco leaves with water, cleaned up by CM-Sephadex column chromatography, and derivatized with fluorescamine (see Chapter 4, Section 1.1). The product is quantitated by ion-pair reversed-phase HPLC and detected fluorometrically. Detection limits of mildiomycin in cucumber, tobacco leaves, and soil were 0.06, 0.2, and 0.7 ppm, respectively, and recoveries were > 70%.

Analysis of mycotoxins in foods by HPLC and other chemical quantification methods were presented in a review by Coker (139).

## REFERENCES

1. H. Nakamura et al., *Anal. Chem.*, **56**, 919 (1984).
2. T. Toyooka et al., *Anal. Chim. Acta*, **149**, 305 (1983).
3. P. Kucera and H. Umagat, *J. Chromatogr.*, **255**, 563 (1983).
4. A. Himuro et al., *J. Chromatogr.*, **264**, 423 (1983).
5. U. A. T. Brinkman et al., *Anal. Chem. Symp. Ser.*, **3**, 247 (1980).
6. J. Yamada et al., *J. Chromatogr.*, **311**, 385 (1984).
7. S. P. Assenza and P. R. Brown, *J. Chromatogr.*, **289**, 355 (1984).
8. H. J. Zeitler et al., *Biochem. Clin. Aspects Pteridines*, **2**, 89 (1983).
9. S. Allenmark et al., *Anal. Biochem.*, **144**, 98 (1985).
10. H. Ezoe et al., *Yamanouchi Seiyaku Kenkyu Hokoku*, **4**, 96 (1980).
11. M. K. Jacobson et al., *Am. J. Physiol.*, **245**, H 887 (1983).
12. A. Ramos-Salazar and A. D. Baines, *Anal. Biochem.*, **145**, 9 (1985).
13. M. Yoshioka et al., *J. Chromatogr.*, **309**, 63 (1984).
14. N. Seiler and B. Knoedgen, *J. Chromatogr.*, **221**, 227 (1980).
15. R. C. Simpson et al., *J. Liq. Chromatogr.*, **5**, 245 (1982).
16. R. F. Minchin and G. R. Hanau, *J. Liq. Chromatogr.*, **7**, 2605 (1984).

17. S. Honda et al., *Anal. Biochem.*, **134**, 483 (1983).
18. S. Honda et al., *J. Chromatogr.*, **281**, 340 (1983).
19. H. Takemoto et al., *Anal. Biochem.*, **145**, 245 (1985).
20. R. V. Lewis, in *CRC Handbook HPLC Sep. Amino Acids, Peptides, Proteins*, Vol. 1, edited by W. S. Hancock, CRC, Boca Raton, Fla., 1984, p. 193.
21. M. Dong et al., *Angew. Chromatogr.*, **42**, 27 (1985).
22. B. R. Larsen and F. G. West, *J. Chromatogr. Sci.*, **19**, 259 (1981).
23. B. N. Jones and J. P. Gilligan, *J. Chromatogr.*, **266**, 471 (1983).
24. H. Godel et al., *J. Chromatogr.*, **297**, 49 (1984).
25. M. H. Joseph and P. Davies, *J. Chromatogr.*, **277**, 125 (1983).
26. H. Liu et al., *Sepu*, **1**, 83 (1984).
27. J. D. H. Cooper et al., *J. Chromatogr.*, **285**, 484 (1984).
28. R. L. Cunico and T. Schlabach, *J. Chromatogr.*, **266**, 461 (1983).
29. M. O. Fleury and D. V. Ashley, *Anal. Biochem.*, **133**, 330 (1983).
30. R. G. Elkin, *J. Agric. Food Chem.*, **32**, 53 (1984).
31. D. L. Hogan et al., *Anal. Biochem.*, **127**, 17 (1982).
32. N. Kondo et al., *Agric. Biol. Chem.*, **48**, 1595 (1984).
33. Z. Deyl and J. Rosmus, *J. Chromatogr.*, **69**, 129 (1972).
34. H. Engelhardt et al., *Anal. Chem.*, **46**, 336 (1974).
35. E. Bayer et al., *Anal. Chem.*, **48**, 1106 (1976).
36. W. McHugh et al., *J. Chromatogr.*, **124**, 376 (1976).
37. Y. Watanabe and K. Imai, *Anal. Biochem.*, **116**, 471 (1981).
38. K. Imai et al., *Chromatographia*, **16**, 214 (1982).
39. Y. Watanabe and K. Imai, *J. Chromatogr.*, **309**, 279 (1984).
40. M. Ahnoff et al., *Anal. Chem.*, **53**, 485 (1981).
41. H. Yoshida et al., *J. High Resolut. Chromatogr. Chromatogr. Commun.*, **5**, 509 (1982).
42. S. Einarsson et al., *J. Chromatogr.*, **282**, 609 (1983).
43. J. J. L'Italien and S. B. H. Kent, *J. Chromatogr.*, **283**, 149 (1984).
44. H. K. Biesalski et al., *GIT Fachz. Lab. Suppl. Chromatogr.*, **4**, 6 (1981).
45. C. A. Collins and C. K. Chow, *J. Chromatogr.*, **317**, 349 (1984).
46. L. A. Morrison and J. A. Driskell, *J. Chromatogr.*, **337**, 249 (1985).
47. A. J. Speek et al., *J. Chromatogr.*, **305**, 53 (1984).
48. J. P. Langenberg and U. R. Tjaden, *J. Chromatogr.*, **305**, 61 (1984).
49. J. Schrijver et al., *Int. J. Vitam. Nutr. Res.*, **51**, 216 (1981).
50. G. M. Murphy et al., *J. Clin. Pathol.*, **23**, 594 (1970).
51. M. Yoshiura et al., *Jikeika Med. J.*, **30**, 207 (1983).
52. Y. Imai et al., *Igaku no Ayumi*, **130**, 212 (1984).

53. D. Ishii et al., *J. Chromatogr.*, **282**, 569 (1983).
54. T. Takeda et al., *Rinsho Kagaku*, **13**, 102 (1984).
55. S. Hasegawa et al., *J. Liq. Chromatogr.*, **7**, 2267 (1984).
56. H. Karatani and S. Oka, *Bunseki Kagaku*, **33**, 6 (1984).
57. E. Reh and G. Schwedt, *Fresenius Z. Anal. Chem.*, **303**, 117 (1980).
58. G. Schwedt and E. Reh, *Chromatographia*, **13**, 779 (1980).
59. T. Kawasaki et al., *J. Chromatogr.*, **226**, 1 (1981).
60. T. Kawasaki et al., *J. Chromatogr.*, **233**, 61 (1982).
61. T. Seki and Y. Yamaguchi, *J. Chromatogr.*, **305**, 188 (1984).
62. H. Lingeman et al., *J. Chromatogr.*, **290**, 215 (1984).
63. T. Sato, *Arch. Jpn. Chir.*, **53**, 33 (1984).
64. H. Tsuchiya et al., *J. Chromatogr.*, **309**, 43 (1984).
65. M. Ikeda et al., *J. Chromatogr.*, **305**, 261 (1984).
66. M. Takeda, *Bunseki Kagaku*, **33**, 681 (1984).
67. F. Sasse et al., *J. Chromatogr.*, **194**, 234 (1980).
68. K. Mopper and D. Delmas, *Anal. Chem.*, **56**, 2557 (1984).
69. T. Toyooka and K. Imai, *J. Chromatogr.*, **282**, 495 (1983).
70. T. Toyooka and K. Imai, *Anal. Chem.*, **56**, 2461 (1984).
71. O. M. Rennert, D. L. Lawson, J. B. Shukla, and T. D. Miale, *Clin. Chim. Acta*, **75**, 365 (1977).
72. J. Turk, S. J. Weiss, J. E. Davis, and P. Needleman, *Prostaglandins*, **16**, 291 (1978).
73. J. F. Cavins and M. Friedman, *J. Biol. Chem.*, **243**, 3357 (1968).
74. M. Minkes, N. Stanford, M. Chi et al., *J. Clin. Invest.*, **59**, 449 (1977).
75. A. M. Krastulovic and A. M. Powell, *J. Chromatogr.*, **171**, 345 (1979).
76. T. Kawasaki, M. Maeda, and A. Tsuji, *J. Chromatogr.*, **163**, 143 (1979).
77. J. T. Taylor, J. G. Knotts, and G. J. Schmidt, *Clin. Chem.*, **26** 130 (1980).
78. K.-G. Wahlund and B. Kdlén, *Clin. Chim. Acta*, **110**, 71 (1981).
79. G. P. Jackman, *Clin. Chim. Acta*, **120**, 137 (1982).
80. G. P. Jackman, *Clin. Chem.*, **26**, 1623 (1980); *ibid.*, **27**, 1202 (1981).
81. A. J. Speek, F. Van Schaik, J. Schrijver, and W. H. P. Schreurs, *J. Chromatogr.*, **228**, 311 (1982).
82. R. S. Rivlin, *Riboflavin*, Plenum Press, New York, London, 1975.
83. J. Marks, *The Vitamins in Health and Disease*, J. and A. Churchill Ltd., London, 1968, p. 85.
84. M. S. Bamji, D. Sharada, and A. N. Naidu, *Int. J. Vitam. Nutr. Res.*, **43**, 351 (1973).
85. H. C. Clarke, *Int. J. Vitam. Nutr. Res.*, **39**, 182 (1969).
86. H. C. Clarke, *Int. J. Vitam. Nutr. Res.*, **47**, 356 (1977).

87. E. Knobloch, R. Hodr, and J. Janda et al., *Int. J. Vitam. Nutr. Res.*, **49**, 144 (1979).
88. H. Mabuchi and H. Nakahashi, *Clin. Chem.*, **29**, 675 (1983).
89. H. Aleyassine and D. B. Tonks, *Clin. Chem.*, **24**, 1849 (1978).
90. H. A. Schwertner and S. B. Hawthorne, *Clin. Chem.*, **26**, 649 (1980).
91. G. Digenis, A. G. Hadjivassiliou, D. Mayopoulou-Symvoulidis et al., *Clin. Chem.*, **27**, 1618 (1981).
92. H. A. Schwertner, *Nephron*, **31**, 209 (1982).
93. J. S. Swan, E. Y. Kragten, and H. Veening, *Clin. Chem.*, **29**, 1082 (1983).
94. J. Goto, M. Saito, T. Chikai et al., *J. Chromatogr.*, **276**, 289 (1983).
95. N. Ichinose, K. Adachi, C. Shimizu et al., *Bunseki Kagaku*, **33**, E 271 (1984).
96. N. Nimura and T. Kinoshita, *Anal. Lett.*, **13**, 191 (1980).
97. S. A. Barker, J. A. Monti, and S. T. Christian et al., *Anal. Biochem.*, **107**, 116 (1980).
98. M. Hatsumi, S. Kimata, and K. Hirosawa, *J. Chromatogr.*, **253**, 271 (1982).
99. S. Imaoka, Y. Fumae, and T. Sugimoto et al., *Anal. Biochem.*, **128**, 459 (1983).
100. Y. Shimomura, K. Taniguchi, and T. Sugie et al., *Rinshokensa*, **27**, 561 (1983).
101. N. Ichinose, K. Nakamura, and C. Shimizu et al., *J. Chromatogr.*, **295**, 463 (1984).
102. T. Watanabe, C. Kitajima, and T. Arakawa et al., *Bull. Jpn. Soc. Sci. Fish.*, **44**, 1109 (1978).
103. T. Watanabe, F. Oowa, C. Kitajima, and S. Fujita, *Bull. Jpn. Soc. Sci. Fish.*, **44**, 1115 (1978).
104. T. Watanabe, T. Arakawa, C. Kitajima et al., *Bull. Jpn. Soc. Sci. Fish.*, **44**, 1223 (1978).
105. T. Takeuchi, T. Watanabe, and T. Nose, *Bull. Jpn. Soc. Sci. Fish.*, **45**, 1319 (1979).
106. T. Hirano and M. Suyama, *Bull. Jpn. Soc. Sci. Fish.*, **49**, 1459 (1983).
107. F. Andreolini, C. Borra, A. D. Corcia et al., *Clin. Chem.*, **30**, 742 (1984).
108. E. Gurpide, M. E. Giebenhain, L. Tseng, and W. G. Kelly, *Am. J. Obstet. Gynecol.*, **109**, 897 (1971).
109. R. J. Liedtke, J. P. Greaves, Jr., J. D. Batjier, and B. Budby, *Clin. Chem.*, **24**, 1100 (1978).
110. J. T. France, B. S. Knox, and P. R. Fisher, *Clin. Chem.*, **28**, 2103 (1982).
111. M. Axelson and J. Sjovall, *J. Steroid Biochem.*, **8**, 683 (1977).
112. P. M. Kabra, F. H. Tsai, and L. J. Marton, *Clin. Chim. Acta*, **128**, 9 (1983).

113. L. A. Kaplan and D. C. Hohnadel, *Clin. Chem.*, **29**, 1436 (1983).
114. A. Di Corcia, R. Samperi, G. Vinci, and G. D'Ascenzo, *Clin. Chem.*, **28**, 1457 (1982).
115. G. Cosmi, A. Di Corcia, R. Samperi, and G. Vinci, *Clin. Chem.*, **29**, 319 (1983).
116. F. Andreolini, A. Di Corcia, A. Laganna et al., *Clin. Chem.*, **29**, 2076 (1983).
117. F. Andreolini, C. Borra, A. Di Corcia et al., *Clin. Chem.*, **31**, 124 (1985).
118. A. L. Barnett and H. Veening, *Clin. Chem.*, **31**, 127 (1985).
119. N. Ichinose, K. Adachi, and G. Schwedt, *Analyst*, **110**, 1505 (1985).
120. G. R. Skurray, *Food Chem.*, **7**, 77 (1981).
121. A. Bognar, *Dtsch. Lebensm. Rundsch.*, **77**, 431 (1981).
122. R. Rouseff, *Liq. Chromatogr. Anal. Food Bev.*, **1**, 161 (1979).
123. A. Henshall, *Liq. Chromatogr. Anal. Food Bev.*, **1**, 31 (1979).
124. C. Y. W. Ang and F. A. Moseley, *J. Agric. Food Chem.*, **28**, 483 (1980).
125. T. Cannella and G. Bichi, *Boll. Chim. Furm.*, **122**, 205 (1983).
126. R. P. Kwok, W. P. Rose, R. Tabor, and T. S. Pattison, *J. Pharm. Sci.*, **70**, 1014 (1981).
127. F. L. Vandemark and G. J. Schmidt, *J. Liq. Chromatogr.*, **4**, 1157 (1981).
128. N. Parris, *J. Agric. Food Chem.*, **32**, 829 (1984).
129. K. E. Claas et al., *Fresenius Z. Anal. Chem.*, **316**, 781 (1983).
130. G. M. Murray and M. J. Sepaniak, *J. Liq. Chromatogr.*, **6**, 931 (1983).
131. H. Ohta et al., *J. Chromatogr.*, **284**, 281 (1984).
132. R. W. Keating and P. R. Haddad, *J. Chromatogr.*, **245**, 249 (1982).
133. J. T. Vanderslice and D. J. Higgs, *J. Chromatogr. Sci.*, **22**, 485 (1984).
134. A. J. Speek et al., *J. Agric. Food Chem.*, **32**, 352 (1984).
135. A. J. Speek et al., *J. Food Sci.*, **50**, 121 (1985).
136. A. J. Speek et al., *J. Chromatogr.*, **301**, 441 (1984).
137. M. Akiyama et al., *Plant Cell Physiol.*, **24**, 1431 (1983).
138. M. Inoue and T. Hagimoto, *Nippon Noyaku Gakkaishi*, **8**, 321 (1983).
139. R. D. Coker, in *Anal. Food Contam.*, edited by J. Gilbert, Elsevier Appl. Sci., London, 1984, p. 207.

# INDEX

ABD-F, fluorometric analysis, with HPLC, 182–183
Abell-Kendall technique, fluorometric measurement of total serum cholesterol, 136–137
Absorption spectra, fluorescence emission characteristics, 13–15
Acetic acid anhydride reagent, derivatization reactions, 94
Acetophenone, resonance structure, 19, 21
Acetyl-spermidine, fluorometric analysis with HPLC, 163–164
Acridinium esters, as immunoassay labels, 128–129
Acridium salts, chemiluminescent analysis of, 33
ADAM (9-anthryldiazomethane):
  EPA levels in fish and plankton, 196–198
  extrinsic fluorophores for carboxylic acid, 29
Adenine nucleotides, fluorometric analysis with HPLC, 162
Adenosine, fluorometric analysis with HPLC, 162
Adrenal cortical steroids:
  fluorescence, 104–106
  structure, 104–105
Adrenaline, derivatization reactions, 80–81
Aequorea, bioluminescence and, 40
ALA-D activity, fluorometric measurements for lead exposure and, 135–136
Albers-Lowry method, cholesterol fluorescence, 104

Alcohol concentrations, fluorescence-photometric enzymatic measurement, 144–146
Alcohol oxidase (AOD), 145–146
Aliphatic amines:
  derivatization reactions, 79
  fluorometric analysis with HPLC, 201, 203
Alkaloids:
  derivatization, 94
  fluorometric analysis with HPLC, 181–182
  native fluorescence, 90–93
  structure of, 89–90
Alkylamines, derivatization reactions, 76
α-fetoprotein (AFP), two-site immunochemiluminometric assay, 128–129
Amines:
  aliphatic:
    derivatization reactions, 79
    fluorometric analysis with HPLC, 201, 203
  aromatic:
    derivatization reactions, 79
    fluorometric analysis with HPLC, 161–163
  biochemistry, 69–83
  biogenic amines, 80–83
  derivatizations, 70–83
  native fluorescence, 69–70
  $NH_2$ compounds, 70–76
  $NH_2$ or $NHR$ compounds, 76–79
  $NR_3$ compounds, 79
  polyamines, 79–80
derivatization reactions, 74

Amines *(Continued)*
  fluorometric analysis:
    with HPLC, 159–160
    volatile, in food chemistry, 201–205
  structural formula, 70
Amino acids:
  derivatization reaction, 72–73, 85–89
  fluorometric analysis with HPLC, 165
  native fluorescence, 83–85
*p*-Aminobenzoic acid, native fluorescence, 96, 98
Aminobutylethyl isoluminol (ABEI):
  chemiluminescent tags, 132
  homogeneous immunoassay, 129–130
9-Aminophananthrene, fluorometric analysis with HPLC, 179–181
Aminophthalate ions, chemiluminescent analysis of, 32–33
Amino sugars, fluorometric analysis with HPLC, 164–165
Ammonia, fluorometric analysis:
  with HPLC, in meat tissue, 201–203
  protein-free blood filtrates, 107–108
Androstenedione, fluorometric analysis with HPLC, 178
Aniline, resonance structures, 19
Aqueous fluorometric continuous-flow method, 142
Aromatic amines:
  derivatization reactions, 79
  fluorometric analysis, with HPLC, 161–163
Aromatic amino acids, native fluorescence, 83–84
Atropine, native fluorescence, 93
Auxochromes, conjugated double bond and resonance structures, 19–21
Avogadro's number, in quantum theory, 6

*Balfourodiniums* salts, native fluorescence, 91–93

Bansyl chloride, derivatization reactions, 76–79
Beer, aliphatic amine fluorescence, with HPLC, 201, 203
*Benecka harveyi* (*Photobacterium fischeri* strain MAV):
  bioluminescence, 37–38
  FMN reductase, 40
  luciferase, 39
  pH profile, 39
  stability, 39
Bile acids, fluorometric analysis of:
  duodenal aspirates, 110–111
  fluorometric evaluation of, 113–114, 193, 196
  with HPLC, 175–177
  serum bile acids, 193, 196
  serum bile acid (SBA), 113–114
Bilirubin, fluorometric analysis of, 138
Biochemistry, luminescence assays in, 29–30
Biogenic amines, derivatization reactions, 80–83
Bioluminescence (BL):
  applications of, 34–41
  defined, 30
  firefly, 36–37
  historical aspects, 30
  marine bacteria, 37–40
  inhibitors, 40
  luciferase, 39
  pH profile, 39
  specificity, 40
  stability, 39
  table of reactions, 35
Biomedical chemistry, fluorometry and, 107–121
  ammonia levels in human plasma, 107–108
  bile acid quantitative analysis, 110
  cholesterol analysis, 107
  diamagnetic cation detection, 117–118
  DNA assays, 115–116
  galactose analysis, 112–113
  inorganic phosphorus (Pi) levels, 115

intracellular pH determination, 116–117
NAD$^+$ assay, 112
near-neutral pH values, 120
pitfalls of ultraviolet light, 119
prolase deficiency measurements, 120–121
serum bile acid analysis, 113–114
tetracycline determination, 108–109
thiobarbituric acid assay, 118–119
unesterified fatty acid levels, 114
uric acid determination, 110–112
uroporphyrinogen I synthase activity, 109–110
Blood plasma:
ammonia concentration in, 107–108
fatty acids in, 180–181
Blue-fluorescent (BFP), aequorea, 40
Bromoacetaldehyde, fluorometric analysis with HPLC, 162

Cadaverine, derivatization reactions, 80
Calibration curves, in fluorometry, 65–66
Carboxylic acids, extrinsic fluorophores for, 24–29
Catecholamines:
derivatization reactions, 80–83
fluorometric analysis:
with HPLC, 188
urinary metanephrine assays, 190–191
Cation-exchange fractionation, imino acid detection, 86
Cellulose acetate electrophoresis, 110–111
Cephaline, native fluorescence, 91–93
Cerebrospinal fluid, fluorometric analysis of:
amino acid analysis, 168–169
polyamines, 186
Chemical structure, fluorescence and, 17–21
Chemiluminescence (CL):
applications, 31–34
acridium salts, 33

diacylhydrazides, 31–33
diaryl oxalates, 33–34
defined, 30
historical aspects, 30
methods, 41
on-line computer analysis of, 133–134
Chemiluminescence energy transfer, 129–130
Chemiluminescence-labeled antibodies, 127–128
Chemiluminescent immunoassay, 133–134
Chemiluminescent tags, in immunoassays, 131–133
Chlorotetracycline:
as fluorescent probe, 117–118
fluorometric determination of, 108–109
*Choisya ternata*, native fluorescence, 91–93
Cholesterol:
fluorometric measurement of:
enzyme estimation of free and total levels, 107
gas-liquid chromatography and, 136–137
structure, 103–104
*Cis* form of planar structure, 18
Cocaine, native fluorescence, 93
Concentration quenching, 16–17
Condensed aromatic hydrocarbons, 19–20
Conjugated double bonds, fluorescence and, 18–21
Corticosterone, fluorometric analysis with HPLC, 178
Cortisol:
fluorometric analysis, with HPLC, 178, 189–190
urinary levels of, luminescent immunoassay for, 133–134
Coumarin, derivatization reactions, 79
Cuvette, fluorometry applications, 58–59
Cyanobenzene, resonance structure, 19, 21

Dansyl chloride (DNS-Cl):
  amino acid detection, thin-layer chromatography, 88–89
  derivatization reactions, 76–79
    fluorometric analysis with HPLC, 170
  structure, 24
Dansyl hydrazine, fluorometric analysis with HPLC, 189–190
Dark electricity, photomultipliers, 61
Dehydroascorbic acid, see Vitamin C
Dehydrogenases, luminescence assays of, 29
Derivatization reactions:
  alkaloids, 94
  amines, 70–83
    biogenic amines, 80–83
    $NH_2$ compounds, 70–76
    NHR compounds, 76–79
    $NR_3$ groups, 79
    polyamines, 79–80
    steps, 74
    steps in, 74
  amino acids, 85–89
    fluorogenic reagent compounds, 87–88
    $o$-phthalaldehyde compounds, 85–87
    thin-layer chromatography, 88–89
  fluorometric analysis with HPLC, 165–170
    $o$-phthalaldehyde, 165–170
    organic acids, 179–181
    postcolumn derivatization, 169–170
    precolumn derivatization, 165–169
  vitamins, 98–101
Diacylhydrazides, chemiluminescence of, 31–32
Diamagnetic cations, fluorometric determination of, 117–118
Diamines, derivatization reactions, 80
Diaryl oxalates, chemiluminescent analysis of, 33–34
Diatomic molecules, photoemission principles and, 9
2,6-Dimethylaminopyridine, derivatization reaction, 75–76
Direct injection technique, fluorometric analysis, with HPLC, 189–190
Direct liquid chromatography, fluorometric analysis with, 199
Direct solid-phase fluoroenzymeimmunoassay, Pd-3G in urine, 133–134
DNA, fluorometric determination of:
  nanogram quantities in cellular homogenates, 115–116
  thiobarbituric acid assay, 118–119
Dopamine, derivatization reactions, 82
Double-beam fluorophotometer, schematic, 54–55
Dual fluorescence differentiation, living and dead cells, 143–145

Early morning urine specimens (EMU), fluorometric analysis of, in normal and infertile women, 140–141
Ecdysteroids, fluorescence, 106
Edman degradation, fluorometric analysis with HPLC, 171–172
Eicosapentaenoic acid (EPA), fluorometric analysis, with HPLC, 196–198
Einstein, Albert, quantum theory, 6
Electron densities, conjugated double bond and resonance structures, 19–20
Electron spin, photoabsorption and emission principles, 7–8
Electron transitions, photoabsorption and emission principles, 7–10
Emetine, native fluorescence, 91–93
Emission, physical properties, 6–12
Emission spectra:
  fluorometry and, 62–64
  light sources and, 53–55
Eosin, planar structure, 18
Estriol, fluorometric analysis, with HPLC:
  adsorption factor, 198–199
  reversed-phase technique, 189

Estriol-16α-glucuronide, fluorometric analysis with HPLC, 199
Estrogen:
  fluorometric analysis of:
    aqueous fluorometric continuous-flow method, 142
    in normal and infertile women, 140–141
    in pregnancy urine, substrate native fluorescence, 139–140
    total estrogens, in pregnancy urine, 142
  native fluorescence, 102–103
  ring-system of, 102–103
  structure, 101–103
Estrogen/creatinine (E/C) ratios, 140–141
Estrone, fluorescence, 103
Ethanethiol, derivatization reaction, 71, 75
Ethylenediamine, derivatization reaction, 82
Excimer lasers, fluorometry applications, 55–56
Excitation spectrum, fluorometry applications, 64–65
Excited state, quantum theory, 7

Fatty acids, fluorometric analysis of:
  with HPLC, 179–181
  EPA levels in fish and plankton, 196–198
  unesterified fatty acid (UFA), 114
Filter fluorometers, 56–57
Fingerprint luminescence, laser detection of, 142–144
Firefly bioluminescence in, 36–37
  assays, 42
Fisheries biochemistry, fluorometric analysis with HPLC, 183
  $B_2$ vitamers, 200–201
  eicosapentaenoic acid levels, 196–198
Flavin adenine dinucleotide (FAD):
  fluorescence emission spectra, 22–23
  fluorometric analysis with HPLC, 192
  fish serum analysis, 200–201
Flavin derivatives, *see specific derivatives, e.g.*: Riboflavin
Flavin mononucleotide (FMN):
  fluorescence emission spectra, 22–23
  fluorometric analysis with HPLC, 192
  fish serum analysis, 200–201
  marine bacteria bioluminescence, 37–38
Flavin mononucleotide (FMN) reductase, 40
2-Fluorenealdehyde, derivatization reactions, 74–75
Fluorescamine:
  derivatization reactions, 70–76
  secondary amino groups, 77–79
  fluorometric analysis with HPLC, 170
Fluorescein:
  planar structure, 18
  structure, 24
Fluorescein diacetate (FDA):
  as pH indicator, 116–117
  simultaneous-flow cytometric DNA analysis, 144
Fluorescein isothiocyanate (FITC), insulin determination, 126
Fluorescence, *see also* Native fluorescence
  calibration curve, 65–66
  chemical structure of molecules, 17–21
    conjugated double bond and resonance structure, 18–21
    planar structure, 18
  divergence of, 58
  fluorophores, 21–29
    extrinsic, 24–29
      carboxylic acids, 24–29
      immunoglobulins, 24
    intrinsic, 21–24
    NADH and $NAD^+$, 24
    proteins, 22
    vitamin $B_2$, 22–23
  inherent emission characteristics, 12–15

Fluorescence, inherent emission
    characteristics *(Continued)*
  Kasha's rule, 12–13
  mirror-image rule, 13–15
  Stokes' shift, 12
 intensity, auxochromes as factor in, 19, 21
 luminescence analysis, 29–43
  biochemical applications, 29–30
  bioluminescence, 34–41
   aequorea, 40
   CL ($H_2O_2$) detection, 41
   firefly, 36–37, 42
   FMN reductase, 40
   marine bacteria, 37–40
  chemiluminescence, 31–34
  historical aspects, 30
  terminology and definitions, 30
 pH, 16
 photoabsorption and emission, 6–12
 photoluminescence, 5–6
 quantum yield, 15
 quenching factors, 16–17
 solvents, 15
 temperature, 16
Fluorescence enhancement immunoassays, 123–124
Fluorescence excitation transfer immunoassay (FETIA), 123–124
Fluorescence-photometric enzymatic measurement, alcohol concentrations, 144–146
Fluorescence polarization immunoassays (FPIA), 123–124
Fluorescence quenching immunoassays, 123–124
Fluorescence stopped-flow technique, 85–87
Fluorescent probes, for fluoroimmunoassays, 123–124
Fluorescent salicylaldehyde-azomethine-diphenylboron chelate:
 derivatization reaction, 76
 food chemistry, 201–205
Fluorogenic reagents, amino acid
    detection with, 87–89
Fluoroimmunoassays:
 applications of, in immunology, 121–122
 types of, 120–124
  heterogeneous, 122
  homogeneous, 122–124
Fluorometry:
 alcohol level measurement, 144–146
 benefits, 1–2
 biochemical applications, 69–106
  alkaloids, 89–94
   derivatization, 94
   native fluorescence, 90–93
  amines, 69–83
   derivatizations, 70–83
   native fluorescence, 69–70
  amino acids and imino acids, 83–89
   derivatization, 85–89
   native fluorescence, 83–85
  steroids, 101–106
   cholesterol, 103–104
   estrogens, 101–103
  vitamins and related compounds, 94–101
   derivatizations, 98–101
   native fluorescence, 96–98
 biomedical and clinical chemistry, 106–146
  biomedical chemistry, 107–121
   ammonia determination in blood plasma, 107–108
   autoanalytical cholesterol estimation, 107
   cellulose acetate electrophoresis, 110–111
   diagmagnetic cation detection, 117–118
   DNA measurement, 115–116
    cytometric analysis, living and dead cells, 143–145
    with thiobarbituric acid assay, 118–119
   galactose analysis, 112–113
   $NAD^+$ luciferase assay, 112
   oxidase assays, 119

pH determination:
  intracellular, 116–117
  near-neutral values, 120
phosphorus analysis, 115
plasma unesterified fatty acid determination, 114
porphyria diagnosis, 109–110
prolidase deficiency analysis, 120–121
serum bile acid analysis in liver disease, 113–114
tetracycline determination, 108–109
uric acid analysis, 110–112
cytometric DNA analysis, 143–145
diagnostic applications, 135–139
  lead exposure indices, 135–136
  Regan enzyme activity in smokers and nonsmokers, 138–139
  saponification efficiency, 136–137
  styrene exposure analysis, 137–138
  tetrapyrrole metabolism, 138
immunologic applications, 121–135
  acridium esters, 128–129
  chemiluminescence energy transfer, 129–130
  chemiluminescence-labeled antibodies, 127–128
  chemiluminescent tags, 131–133
  cortisol analysis, 133–134
  heterogeneous fluoroimmunoassays, 122
  homogeneous fluoroimmunoassays, 122–124
  laser immunoassay of insulin, 125–126
  luminescence immunoassay of human serum albumin, 126–127
  Pd-3G urine analysis, 133–134
  peroxidase-labeled conjugates, 130–131
  progesterone analysis, 134–135
  two-site immunochemi-luminometric assay, 129
inherent fingerprint luminescence, 142–144

measurement principles, 51–66
  apparatus and arrangements, 51–61
  calibration curve, 65–66
  emission spectrum, 62–64
  excitation spectrum, 64–65
  light sources, 53–56
  measuring cells, 58–60
  monochromator and filters, 56–57
  photomultiplier, 60–61
obstetrics and gynecology, 139–146
  estrogen analysis, 139–140
  urine analysis of estrogen levels:
    aqueous fluorometric continuous-flow method, 142
    fertile and nonfertile women, 140–141
    total estrogen analysis, 142
Fluorophores, 21–29
  cholesterol determination, 107
  fluorescence emission characteristics, 14–15
  intrinsic, 21–24
    NADH and $NAD^+$, 24
    proteins, 22
    vitamin $B_2$, 22–23
Folic acid (pteroylglutamic acid), derivatization reaction, 101
Food chemistry, fluorometric analysis with HPLC, 201–205
"Forbidden transitions," photoabsorption and emission principles, 7–8
Frank–Condon principle, 8
"Front-face" fluorometry, tetrapyrrole metabolism, 138
Fused silicon dioxide, fluorometry measuring cells, 58

Galactosamine, fluorometric analysis with HPLC, 164–165
Galactose, fluorometric analysis of, 112–113
Galactosemia, fluorometric diagnosis of, 112–113

# 218　　INDEX

Galactosuria, fluorometric diagnosis of, 112–113
Gallic acid, chemiluminescent analysis of, 34
Gas lasers, fluorometry applications, 55
Gas-liquid chromatography, fluorometric measurement and, total serum cholesterol, 136–137
Gastric juices, fluorometric analysis with HPLC, 168–169
Glucosamine, fluorometric analysis with HPLC, 164–165
Glutamate dehydrogenase (GDH), fluorometric measurement, 108
GOT-GPT, fluorometric evaluation of serum bile acid (SBA), 113–114

Harmane alkaloids, fluorometric analysis with HPLC, 181–182
Heat-stable alkaline phosphatase (HSAP), 138–139
Hemin catalyst immunoassay, human serum albumin, 126–127
Hemodialysate solutions, fluorometric analysis with HPLC, 199–200
Heterogeneous fluoroimmunoassays, 120
High-performance liquid chromatography (HPLC), fluorometric analysis and, 1–2, 159–183
  alkaloids, 181–182
  amines (general), 159–160
  amino acids and imino acids, 165
  amino sugars, 164–165
  $B_2$ vitamers:
    in fish serum, 200–201
    flavin-adenine dinucleotide levels, 192
  cerebrospinal fluid polyamine monitoring, 186
  conditions, 183–185
  cortisol levels with dansyl hydrazine, 189–190
  derivatization reactions and, 83, 165–172
  postcolumn derivatization, 169–170
  precolumn derivatization, 165–169
  eicosapentaenoic acid, 196–198
  estriol-16α-glucouronide, 199
  food chemistry and, 201–205
  fluorescent materials in uremic fluids, 193, 195
  hemodialysate solutions, 199–200
  $N$-heterocyclic compounds, 160–163
  5-hydroxyindole-3-acetic acid levels in urine, 189–190
  measuring cells, 59–60
  organic acids, 179–181
  placental estriol levels, 189
  polyamines, 163–164
  prostaglandins and thromboxanes derivatives, 186–188
  sera and urine levels, uremic and normal subjects, 192–193
  serum bile acids, 193, 196
  steroids, 175–178
    bile acids, 175–177
  stopped-flow scanning technique for catecholamines, 188
  thiols, 182–183
  unconjugated estriol in maternal serum/plasma, 198–199
  urinary metanephrines, 190–191
  vitamins, 172–175
Histamine, derivatization reactions, 80
HOE 33662 compound, simultaneous-flow cytometric DNA analysis, 144
Homogeneous fluoroimmunoassays, 122–124
  chemiluminescence energy transfer, 129–130
Homovanillic acid, fluorometric oxidase assays, 119
Horseradish peroxidase (HRP):
  enhanced luminescence procedure, 130–131
  fluorometric oxidase assays, 119
HPTS, as fluorescent pH indicator, 120
Human plasma, fluorometric analysis,

with HPLC, 168–169
Human serum albumin, luminescence immunoassay, 126–127
Human skin fibroblast prolidase, fluorometric determination of, 120–121
Hydrogen peroxide, chemiluminescent analysis and, 32–33
5-Hydroxyindole-3-acetic acid, fluorometric analysis with HPLC, 189–190
Hydroxycarbonium ion, fluorometric analysis, 137–138
17-Hydroxy corticosteroids, fluorometric analysis with HPLC, 178
$p$-Hydroxyphenylacetic acid fluorophor, 110–112
Hydroxyproline, fluorometric analysis with HPLC, 169
3α-Hydroxysteroid dehydrogenase, bile acid analysis, 110–111
Hyoscine, native fluorescence, 93

Imino acids:
  cation-exchange fractionation, 86
  fluorometric analysis:
    with HPLC, 165
    $o$-phthalaldehyde, 165–166
    native fluorescence, 83–85
Immunofluorometric assays (IFMA), 120
Immunology:
  fluorometry and, 121–135
    acridinium esters, 128–129
    chemiluminescence energy transfer, 129–130
    chemiluminescence-labeled antibodies, 127–128
    chemiluminescent tags, 131–133
    direct-phase fluoroenzyme-immunoassay, 133–134
    enhanced luminescence, 130–131
    fluoroimmunoassays, 122–124
      heterogeneous assays, 122
      homogeneous assays, 122–124
    laser fluorescence immunoassays, 125–126

luminescence immunoassays, 126–127
on-line computer analysis, 133–134
solid-phase chemiluminescence immunoassay, 134–135
two-site immunochemiluminometric assay, 129
Immunoradiometric assays (IRMA), 120
Indoleacetic acid, fluorometric analysis with HPLC, 205
Indoles, native fluorescence, 69
Inhibitors:
  firefly bioluminescence and, 37
  marine bacteria bioluminescence, 40
Insulin, laser fluorescence immunoassay, 125–126
Intercrossing system, 11–12
Internal conversion, 11
Ion-exchange chromatography:
  amino acid detection, $o$-phthalaldehyde, 85–87
  fluorometric analysis, with HPLC, 167–168
Ipecacuanha alkaloids, native fluorescence, 91–93
Isoalloxazine, structure, 22–23
Isothiocyanates, structure, 24

Kasha's Rule, fluorescence emission characteristics, 12–13
"Key atoms," conjugated double bond and resonance structures, 19–20
Kober-Ittrich reaction, estrogen fluorescence, 102–103
Kober's reagent, estrogen fluorescence, 102–103

Laser fluorescence immunoassay of insulin, 125–126
Lasers:
  fingerprint luminescence detection, 142–144
  fluorometry applications, 55–56

Lead exposure, fluorometric
    measurements for, 135–136
Leukemia, cerebrospinal fluid
    monitoring, fluorometric analysis
    with HPLC, 186
Light sources, fluorometry, 53–56
Lipids, fluorometric oxidase assays,
    119
Liquid lasers, fluorometry
    applications, 55
Liver disease, fluorometric evaluation
    of serum bile acid (SBA), 113–114
Lophine, chemiluminescent analysis
    of, 34
Luciferase:
    firefly bioluminescence and, 37
    marine bacteria bioluminescence,
        39
    NAD$^+$, 112
Lucigenin, chemiluminescent analysis
    of, 33
Lumichrome (LC), RF fluorescence
    and, 23
Lumiflavin:
    native fluorescence, 96–97
    photochemical conversion from
        riboflavin, 99–100
    RF fluorescence and, 23
Luminescence analysis:
    biochemical applications, 29–30
    bioluminescence, 34–41
        aqueorea, 40
        firefly, 38–39, 42
        marine bacteria, 37–40, 43
    chemiluminescence, 31–34
        acridinium salts, 33
        diacylhydrazides, 31–33
        diaryl oxalates, 33–34
    historical aspects, 30
    immunoassay, human serum
        albumin, 126–127
    terminology and definitions, 30
Luminogenic compound-labeled
    antigens, 131–133
Luminol, chemiluminescent analysis
    of, 33

Malonic acid, derivatization reactions,
    94
Mandelic acids, fluorometric
    measurement of, 137–138
Marine bacteria:
    bioluminescence, 37–40
    assays, 43
    inhibitors, 40
    luciferase, 39
    pH profile, 39
    specificity, 40
    stability, 39
MDPF (2-methoxy-2,4-diphenyl-
    3(2$H$)furanone):
    derivatization reactions, 77–79
    fluorometric analysis with HPLC,
        159–160
Measuring cells, fluorometry, 58–60
Medical diagnoses, fluorometric
    measurements for, 135–139
    cerebrospinal fluid polyamine
        monitoring, 186
    hemodialysate solutions, 199–200
    5-hydroxyindole-3-acetic acid levels
        in urine, 189–190
    insulin, laser fluorescence
        immunoassay, 125–126
    lead exposure index, 135–136
    liver disease, serum bile acid
        evaluation, 113–114
    placental estriol levels, 189
    prostaglandins and thromboxanes
        derivatives, 186–188
    Regan enzyme activity, 138–139
    saponification efficiency, 136–137
    sera and urine levels, uremic and
        normal subjects, 192–193
    serum bile acids, 193, 196
    steroids, 175–178
        bile acids, 175–177
    styrene exposure, 137–138
    tetrapyrrole metabolism, 138
    urinary metanephrines, 190–191
β-Mercaptoethanol, derivatization
    reaction, 71, 75
3-Mercaptopropionic acid,

fluorometric analysis with HPLC, 160
Mercury-cadmium lamp, fluorometric measurement, 55
Mercury vapor lamps, emission spectra, 54
Mercury-xenon, and fluorometry, 55
Metanephrines (urinary), fluorometric analysis with HPLC, 190–191
o-Methylmetabolites, derivatization reactions, 81–82
Micellar systems, fluorometric determination of amino acids, 87–88
Mildiomycin residues, fluorometric analysis, with HPLC, food chemistry and, 205
Mirror-image rule, fluorescence emission characteristics, 13–14
Monoamine fluorophores, derivatization reactions, 82–83
Monochromator:
  fluorometry applications, 56–57
  grating apparatus, 52
Morphine, native fluorescence, 93
Multiplicity, photoabsorption and emission principles, 8
Mycotoxins, fluorometric analysis with HPLC, 205

NaClO, derivatization reactions, 78–79
NAD, fluorometric analysis with HPLC, bile acids, 176
$NAD^+$:
  ammonia concentration in blood plasma, 107–108
  fluorescence emission spectra, 24
  luciferase assay, 112
NADH (nicotinamide adenine dinucleotide):
  ammonia concentration in blood plasma, 107–108
  fluorescence emission spectra, 24
  fluorometric analysis with HPLC, 176

  structure, 108
$NADH_2$, marine bacteria bioluminescence, 38–39
NADPH, Pd-3G levels in urine, 133–134
Native fluorescence:
  adrenal cortical steroids, 105–106
  alkaloids, 90–93
  amines, 69–70
  amino acids and imino acids, 83–85
  cholesterol, 104
  estrogen, 102–103
  fluorometric analysis:
    in pregnancy urine, 139–140
  with HPLC:
    stopped-flow scanning techniques, 188
    uremic patients, 193–195
NBC-chloride, fluorometric analysis with HPLC, 203
NBD-chloride:
  derivatization reactions, 76–79
  fluorometric analysis with HPLC, 171
NBD-F, fluorometric analysis with HPLC, 159–160
NBD-fluoride, fluorometric analysis with HPLC, 170–171
$NH_2$ groups:
  derivatization reactions, 70–76
  NHR groups and, 76–79
N-Heterocyclic compounds, fluorometric analysis, 160–163
Nicotinamide, derivatization reaction, 98–100
Norepinephrine, derivatization reactions, 82
$NR_3$ groups, derivatization reactions, 79
Nucleosil-$NO_2$, fluorometric analysis with HPLC, 177–178

Obstetrics and gynecology, fluorometric analysis and, 139–146
  aqueous fluorometric continuous-

Obstetrics and gynecology,
    fluorometric analysis and
    *(Continued)*
  flow method, 142
  estrogen levels:
    in normal and infertile women,
      140–141
    in pregnancy urine, 139–140
    total estrogen levels in pregnancy
      urine, 142
Off-plane spectrophotosystem, 59–60
Opium alkaloids, native fluorescence,
  93
Organic acids, fluorometric analysis
  with HPLC, 179–181
Organic solvents, fluorescence
  intensities, 83–84
Oxidases, luminescence assays of, 29
17-Oxosteroids, fluorometric analysis
  with HPLC, 178
Oxytetracycline, fluorometric
  determination of, 108–109

Paper chromatography, estrogen
  fluorescence, 103
Pd-3G, urinary levels of, 133–134
Peptides, fluorogenic determination,
  87
Permeability, filter fluorometers, 57
Peroxidase-labeled conjugates,
  enhanced luminescence
  procedure, 130–131
Peroxidases, chemiluminescent
  analysis of, 33
Perylene, fluorescence emission
  characteristics, 14–15
pH:
  fluorescence and, 16
    intracellular measurements,
      116–117
    near-neutral values, 120
  marine bacteria bioluminescence,
    39
*o*-Phthalaldehyde:
  derivatization reactions, 85–87
  fluorometric analysis with HPLC,
    159–160

continuous-flow method, 162
  derivatization reactions, 165–170
Phenol, resonance structures, 19
Phenylglyoxylic acid, styrene exposure
  and, 137–138
Phenylthiohydantoin amino acids,
  88–89
Phosphorescence, physical properties,
  5–6
Phosphorus (inorganic) (Pi), 115
Photoabsorption, physical properties,
  6–12
Photoluminescence, physical
  properties, 5–6
Photomultiplier, 60–61
Photoproteins, aequorea, 40
PITC reagent, fluorometric analysis
  with HPLC, 171–172
Planar structure, fluorescence and, 18
Planck's constant, in quantum theory,
  6
*Platydesminium* salts, native
  fluorescence, 91–93
Polyamines:
  cerebrospinal fluid:
    fluorometric analysis with HPLC,
      186
  derivatization reactions, 79–80
  fluorometric analysis with HPLC,
    163–164
Porphobilinogen (PBG):
  lead exposure index, 135–136
  uroporphyrinogen I synthase
    activity, 109–110
Porphyria, fluorometric determination
  of, 109–110
Porphyrins, tetrapyrrole metabolism,
  138
Progesterone, chemiluminescence
  immunoassay for, 134–135
Prolidase deficiency, fluorometric
  determination of, 120–121
Proline, fluorometric analysis, with
  HPLC, 169
Prostaglandins, fluorometric analysis
  with HPLC, 186–188
Protein microsequencing, fluorometric

analysis with HPLC, 171–172
Purines, fluorometric analysis with HPLC, 161–163
Putrescine:
  derivatization reactions, 80
  fluorometric analysis with HPLC, 163–164
1-Pyrenealdehyde, derivatizations, 74–75
Pyrene butyric acid, as oxygen indicator, 145–146
Pyrex glass, fluorometry measuring cells, 58
Pyridoxal, native fluorescence, 96–98
Pyridoxamine, native fluorescence, 96–98
Pyridoxine:
  fingerprint luminescence detection, 143
  native fluorescence, 96–98
Pyrimidines, fluorometric analysis with HPLC, 161–163
Pyrogallol, chemiluminescent analysis of, 34

Quantum mechanics, photoabsorption and emission principles, 6–12
Quartz, fluorometry measuring cells, 58
Quenching, fluorescence and, 16–17

Radioimmunoassays, limits of, in immunology, 121
Raman spectrum, fluorescence emission spectrum, 62–63
Rayleigh scattering, fluorescence emission spectrum, 62–63
Receiver character, photomultipliers, 61
Regan enzyme, fluorometric analysis of, 138–139
Release fluoroimmunoassays, 123–124
Renal tubular fluid, 115
Resazurin, formation of resorfin from, 113–114
Resonance radiation, 8–9
Resonance structure, fluorescence and, 8–9, 18–21
Resorfin, formation of, from resazurin, 113–114
Reversed-phase HPLC, urinary placental estriol, 189
Rhodamine isocyanates, structure, 24
Riboflavin:
  derivatization reaction, 99–100
  fingerprint luminescence detection, 143
  fluorescence emission spectra, 22–23
  fluorometric analysis with HPLC, 192
  fish serum analysis, 200–201
  native fluorescence, 96–97

Salicylic acid, fluorescence emission spectrum, 62
Saponification efficiency, total serum cholesterol, 136–137
SBD-F, fluorometric analysis with HPLC, 182–183
Schiff's bases, derivatization reactions, 74–75
"Sensitized chemiluminescence," 31
Separation fluoroimmunoassays (Sep-FIAs), 120
Serotonin, derivatization reactions, 82
Serum bile acid (SBA), fluorometric evaluation of, 113–114
  with HPLC, 193, 196
Siloxine, chemiluminescent analysis of, 34
Silver/silver bromide, photomultipliers, 60–61
Simultaneous-flow cytometric DNA analysis, 143–145
Sodium salicylate, excitation spectrum, 64–65
Solid lasers, fluorometry applications, 55
Solid-phase antigen luminescence technique (SPALT), 132–133
Solid-phase chemiluminescent immunoassay, 134–135
Solvents:
  fluorescence and, 15

Solvents, fluorescence and *(Continued)*
    amino acids, 83–84
    Raman-scattering radiation, 63–64
Specificity:
    aequorea, 40
    firefly bioluminescence and, 37
    marine bacteria bioluminescence, 40
Spectrofluorometer, apparatus and arrangement of, 51–52
Spectrophotofluorometer, optic system, 58–59
Spermidine:
    derivatization reactions, 80
    fluorometric analysis with HPLC, 163–164
Spermine:
    derivatization reactions, 80
    fluorometric analysis with HPLC, 163–164
Spin angular momentum, photoabsorption and emission principles, 7–8
Stability, marine bacteria bioluminescence, 39
Steric hindrance, derivatization reaction, 71
Steroids:
    fluorescence, 101–106
        adrenal cortical steroids, 104–106
        cholesterol, 103–104
        estrogens, 101–103
    fluorometric analysis with HPLC, 175–178
    bile acids, 175–177
Stokes law, fluorescence emission spectrum, 62–63
Stokes' shift, fluorescence emission characteristics, 12
Stopped-flow scanning technique, 188
Styrene exposure, mandelic and phenylglyoxylic acids, 137–138

TCPO (bis(trichlorophenyl) oxalate), 33–34
Temperature, fluorescence and, 16
Tetracyclines, fluorometric determination of, 108–109
Tetrapyrrole metabolism, fluorometric analysis of, 138
Thiamin:
    derivatization reactions, 98–99
    fluorometric analysis:
        inorganic phosphorus and thiochrome conversion, 115–116
        with HPLC, food chemistry and, 203
    native fluorescence, 96–97
Thin-layer chromatography (TLC):
    alkaloid detection, 94
    amino acid detection, 88–89
Thiobarbituric acid assay, fluorometric determination of DNA, 118–119
Thiochrome:
    derivatization of thiamine, 98–99
    fluorometric determination of inorganic phosphorus and, 115–116
    native fluorescence, 96–98
2,2-Thiodiethanol, derivatization reactions, 78–79
Thiols, fluorometric analysis, with HPLC, 182–183
Thromboxanes, fluorometric analysis, with HPLC, 186–188
Tocopherols:
    fluorometric analysis, with HPLC, 204
    native fluorescence, 96, 98
*Trans* form of planar structure, 18
Trihydroxyindole compounds, derivatization reactions, 80–81
Tryptamine, derivatization reactions, 82
Tryptophans:
    fluorometric analysis with HPLC, 188
    native fluorescence, 83–84
Tungsten lamp:
    fluorometry applications, 51–52
    light source, 53
Two-site immunochemiluminometric assay, 129

Two-site luminometric assay, 128
Tyrosine, native fluorescence, 83–84

Ultraviolet light, fluorometric oxidase assays, 119
Unesterified fatty acid (UFA), 114
Uremic patients, fluorometric analysis with HPLC:
  reversed-phase techniques, 193, 195
  sera and urine fluorescence, 192–194
Uric acid, fluorometric determination of, 110–112
Uronic acids, fluorometric analysis with HPLC, 180–181
Uroporphyrin, structure of, 135–136
Uroporphyrinogen I synthase, fluorometric determination of, 109–110

*Vibrio fischeri*:
  bioluminescence, 37–38
  inhibitors, 40
  luciferase, 39
  pH profile, 39
  specificity, 40
  stability, 39
Vitamin A (retinol):
  fluorometric analysis with HPLC, 174
  native fluorescence, 96–97
Vitamin $B_1$, *see* Thiamin
Vitamin $B_2$, *see* Riboflavin
Vitamin $B_6$, fluorometric analysis with HPLC, 174
Vitamin $B_{12}$, native fluorescence, 98
Vitamin C:
  derivatization reaction, 99
  fluorometric analysis with HPLC, 174–175
Vitamin $D_2$, derivatization reaction, 99–100
Vitamin $D_3$, derivatization reaction, 99–100
Vitamin E, *see* Tocopherols
Vitamin $K_1$ and $K_2$, fluorometric analysis with HPLC, 175, 204–205
Vitamins:
  derivatizations, 98–101
  fluorometric analysis with HPLC, 172–175
  food chemistry and, 203–204
  native fluorescence, 96–98
  structure, 94–96

Walker ascites tumor, simultaneous-flow cytometric DNA analysis, 143–144

Xenon lamps:
  emission spectrum, 53–54
  fluorometry applications, 51–52

Yellow enzymes, structure, 22–23
Yellow-green fluorescence, estrogens, 102

325300

QP     Fluorometric Analysis in
519.9   Biomedical Chemistry
.F58
F58
1991

| DATE DUE | | | |
|---|---|---|---|
| 5/5/93 | | | |
| | | | |
| | | | |
| | | | |
| | | | |
| | | | |
| | | | |
| | | | |
| | | | |
| | | | |
| | | | |
| | | | |

**Lebanon Valley College Library**

Annville, Pennsylvania   **17003**

(*continued from front*)

Vol. 63. **Applied Electron Spectroscopy for Chemical Analysis.** Edited by Hassan Windawi and Floyd Ho

Vol. 64. **Analytical Aspects of Environmental Chemistry.** Edited by David F. S. Natusch and Philip K. Hopke

Vol. 65. **The Interpretation of Analytical Chemical Data by the Use of Cluster Analysis.** By D. Luc Massart and Leonard Kaufman

Vol. 66. **Solid Phase Biochemistry: Analytical and Synthetic Aspects.** Edited by William H. Scouten

Vol. 67. **An Introduction to Photoelectron Spectroscopy.** By Pradip K. Ghosh

Vol. 68. **Room Temperature Phosphorimetry for Chemical Analysis.** By Tuan Vo-Dinh

Vol. 69. **Potentiometry and Potentiometric Titrations.** By E. P. Serjeant

Vol. 70. **Design and Application of Process Analyzer Systems.** By Paul E. Mix

Vol. 71. **Analysis of Organic and Biological Surfaces.** Edited by Patrick Echlin

Vol. 72. **Small Bore Liquid Chromatography Columns: Their Properties and Uses.** Edited by Raymond P. W. Scott

Vol. 73. **Modern Methods of Particle Size Analysis.** Edited by Howard G. Barth

Vol. 74. **Auger Electron Spectroscopy.** By Michael Thompson, M. D. Baker, Alec Christie, and J. F. Tyson

Vol. 75. **Spot Test Analysis: Clinical, Environmental, Forensic and Geochemical Applications.** By Ervin Jungreis

Vol. 76. **Receptor Modeling in Environmental Chemistry.** By Philip K. Hopke

Vol. 77. **Molecular Luminescence Spectroscopy: Methods and Applications** (*in two parts*). Edited by Stephen G. Schulman

Vol. 78. **Inorganic Chromatographic Analysis.** Edited by John C. MacDonald

Vol. 79. **Analytical Solution Calorimetry.** Edited by J. K. Grime

Vol. 80. **Selected Methods of Trace Metal Analysis: Biological and Environmental Samples.** By Jon C. VanLoon

Vol. 81. **The Analysis of Extraterrestrial Materials.** By Isidore Adler

Vol. 82. **Chemometrics.** By Muhammad A. Sharaf, Deborah L. Illman, and Bruce R. Kowalski

Vol. 83. **Fourier Transform Infrared Spectrometry.** By Peter R. Griffiths and James A. de Haseth

Vol. 84. **Trace Analysis: Spectroscopic Methods for Molecules.** Edited by Gary Christian and James B. Callis

Vol. 85. **Ultratrace Analysis of Pharmaceuticals and Other Compounds of Interest.** Edited by S. Ahuja

Vol. 86. **Secondary Ion Mass Spectrometry: Basic Concepts, Instrumental Aspects, Applications and Trends.** By A. Benninghoven, F. G. Rüdenauer, and H. W. Werner

Vol. 87. **Analytical Applications of Lasers.** Edited by Edward H. Piepmeier

Vol. 88. **Applied Geochemical Analysis.** by C. O. Ingamells and F. F. Pitard

Vol. 89. **Detectors for Liquid Chromatography.** Edited by Edward S. Yeung

Vol. 90. **Inductively Coupled Plasma Emission Spectroscopy: Part I: Methodology, Instrumentation, and Performance; Part II: Applications and Fundamentals.** Edited by J. M. Boumans

Vol. 91. **Applications of New Mass Spectrometry Techniques in Pesticide Chemistry.** Edited by Joseph Rosen